AF588103

Springer Theses

Recognizing Outstanding Ph.D. Research

For further volumes:
http://www.springer.com/series/8790

Aims and Scope

The series "Springer Theses" brings together a selection of the very best Ph.D. theses from around the world and across the physical sciences. Nominated and endorsed by two recognized specialists, each published volume has been selected for its scientific excellence and the high impact of its contents for the pertinent field of research. For greater accessibility to non-specialists, the published versions include an extended introduction, as well as a foreword by the student's supervisor explaining the special relevance of the work for the field. As a whole, the series will provide a valuable resource both for newcomers to the research fields described, and for other scientists seeking detailed background information on special questions. Finally, it provides an accredited documentation of the valuable contributions made by today's younger generation of scientists.

Theses are accepted into the series by invited nomination only and must fulfill all of the following criteria

- They must be written in good English.
- The topic should fall within the confines of Chemistry, Physics, Earth Sciences, Engineering and related interdisciplinary fields such as Materials, Nanoscience, Chemical Engineering, Complex Systems and Biophysics.
- The work reported in the thesis must represent a significant scientific advance.
- If the thesis includes previously published material, permission to reproduce this must be gained from the respective copyright holder.
- They must have been examined and passed during the 12 months prior to nomination.
- Each thesis should include a foreword by the supervisor outlining the significance of its content.
- The theses should have a clearly defined structure including an introduction accessible to scientists not expert in that particular field.

Takashi Kumagai

Visualization of Hydrogen-Bond Dynamics

Water-Based Model Systems on a Cu(110) Surface

Doctoral Thesis accepted by
Kyoto University, Japan

Author
Dr. Takashi Kumagai
Fritz-Haber Institute of the Max-Planck Society
Berlin
Germany

Supervisor
Prof. Hiroshi Okuyama
Kyoto University
Kyoto
Japan

ISSN 2190-5053 ISSN 2190-5061 (electronic)
ISBN 978-4-431-54155-4 ISBN 978-4-431-54156-1 (eBook)
DOI 10.1007/978-4-431-54156-1
Springer Tokyo Heidelberg New York Dordrecht London

Library of Congress Control Number: 2012941634

Printed on acid-free paper

Springer is part of Springer Science+Business Media (www.springer.com)

Parts of this thesis have been published in the following journal articles:

T. Kumagai, M. Kaizu, S. Hatta, H. Okuyama, T, Aruga. I. Hamada, Y. Morikawa, *Direct Observation of Hydrogen-Bond Exchange within a Single Water Dimer*, Physical Review Letters 100 (2008) 166101. Copyright American Physical Society. Reproduced with permission.

T. Kumagai, M. Kaizu, H. Okuyama, S. Hatta, T. Aruga, I. Hamada, Y. Morikawa, *Water Monomer and Dimer on Cu(110) Studied Using a Scanning Tunneling Microscope*, e-Journal of Surface Science and Nanotechnology, 6 (2008) 296. Copyright The Surface Science Society of Japan. Reproduced with permission.

T. Kumagai, M. Kaizu, H. Okuyama, S. Hatta, T. Aruga, I. Hamada, Y. Morikawa, *Tunneling dynamics of a hydroxyl group adsorbed on Cu(110)*, Physical Review B 79 (2009) 035423. Copyright American Physical Society. Reproduced with permission.

T. Kumagai, M. Kaizu, H. Okuyama, S. Hatta, T. Aruga, I. Hamada, Y. Morikawa, *Symmetric hydrogen bond in a water-hydroxyl complex on Cu(110)*, Physical Review B 81 (2010) 045402. Copyright American Physical Society. Reproduced with permission.

T. Kumagai, H. Okuyama, S. Hatta, T. Aruga, I. Hamada, *Water clusters on Cu(110): chain versus cyclic structures*, Journal of Chemical Physics 134 (2011) 024703. Copyright American Institute of Physics. Reproduced with permission.

T. Kumagai, A. Shiotari, H. Okuyama, S. Hatta, T. Aruga, I. Hamada, T. Frederiksen, H. Ueba, *H-atom relay reactions in real space*, Nature Materials, 11 (2012) 167. Copyright Nature Publishing Group. Reproduced with permission.

Supervisor's Foreword

The water molecule, H_2O, is one of the most familiar molecules around us. Its unique ability to develop an H-bond network plays a central role in so many areas of chemistry and biology, yet it remains poorly understood at the microscopic level. Furthermore, H-bond dynamics, such as bond rearrangements and proton transfer, are difficult to probe and researchers have invariably relied on spectroscopic probes to infer H motions. In his Ph.D. work, Takashi Kumagai achieved a direct probe of the H-bond dynamics on a metal surface using a scanning tunneling microscope (STM). By combining its manipulation capabilities with single-molecule imaging, he was able to construct a variety of H-bond structures on the surface. Properties and dynamics of individual H-bonds in water clusters, hydroxyl clusters, and water-hydroxyl complexes were studied in conjunction with density functional theory. Most notably, concerted proton relay reactions, which are frequently invoked across many fields of chemistry, were visualized and controlled by tunneling electrons. One of the key aspects that distinguishes this work from others is the rational choice of a Cu(110) surface as the substrate. Because water molecules interact weakly with copper surfaces, it was not feasible to controllably manipulate a water molecule with the STM on low-index surfaces, such as Cu(111) and (100). The relatively "open" structure of the (110) plane makes the outermost Cu atoms electron-deficient, which causes water molecules to be moderately bound on the surface. Furthermore, the 2-fold symmetry of the substrate makes the dynamics of water and its derivatives anisotropic, which allows us to investigate them in a straightforward way. Takashi Kumagai's great efforts and patience that always must accompany STM experiments were certainly essential. This thesis paves the way for studying H-bond dynamics in real space, providing an important contribution to fields beyond surface chemistry.

Hiroshi Okuyama

Acknowledgments

First, I would like to sincerely thank Prof. Hiroshi Okuyama of Kyoto University, Japan, the director of my Ph.D. program, for giving me great opportunities to work in a fascinating research environment. While supervising me, he treated me as a colleague, engaged me in challenging discussions, and always motivated me to push my research forward.

I also sincerely appreciate Prof. Tetsuya Aruga, Prof. Shinichiro Hatta, and other members of the laboratory at Kyoto University for the fruitful discussions I had with them in the completion of my work.

I am grateful to Mr. Masahisa Kaizu at Kyoto University; Dr. Ikutaro Hamada at Tohoku University, Japan; Prof. Yoshitada Morikawa at Osaka University; Dr. Thomas Frederiksen at the Donostia International Physics Center, Spain; and Prof. Hiromu Ueba at Toyama University for collaborating in theoretical works.

I also thank the following individuals: Dr. Sebastian Stepanow and Prof. Klaus Kern at the Max-Planck Institute at Stuttgart, Germany, for accommodating me as a visiting student for half a year; Dr. Kenta Motobayashi and Mr. Tomonari Okada at RIKEN, Japan, for sharing their Ph.D. and master's theses, which motivated and inspired me to summarize this thesis; Mr. Yuichi Fujimori, Dr. Christoph Nacci, and other members of the Fritz-Haber Institute of the Max-Planck Society for helping with the proof of my thesis; and Mr. Hideo Morishima.

My deep gratitude goes to the Japan Society for the Promotion of Science for the scholarship received during the period of my Ph.D. coursework.

Finally, I am grateful to my family for their support.

Takashi Kumagai

Acknowledgments

Contents

Chapter 1
Introduction

Abstract The main topic of this thesis is visualization of H-bond dynamics within water-based model systems assembled on a metal surface. Scanning tunneling microscope (STM) is employed to image and engineer such model systems at the single molecule level. H-bond dynamics of water molecules, i.e., the bond exchange, H-atom transfer reactions, are a pretty simple and important process in chemistry and biology, which also deeply relates to fundamentals of life activities. It is usually very difficult to investigate H-bond dynamics in chemical and biological systems at the single molecule level because complicated environmental effects cannot be eliminated in conventional spectroscopies in which the signal stems from the ensemble of the system. In this thesis I show the characterization of H-bonding nature and dynamics within water-based model systems at the spatial limit. Experiments under ultra-high vacuum and low temperature conditions eliminate the environmental effects and thermal fluctuations. General discussions and previous studies of H-bond dynamics of water are described in Sect. 1.1. Previous studies of water on metal surfaces are summarized in Sect. 1.2. The evolution of single atom/molecule science using STM is described in Sect. 1.3. Finally, I mention the motivation and scope in Sect. 1.4.

Keywords Hydrogen-bond dynamics · Water molecules on metal surfaces · Single atom/molecule science · Scanning tunneling microscope

1.1 Dynamics of Hydrogen Bond in Water

H bond is a moderate interaction between molecules. The strength of H bond is larger than that of van-der-Waals interaction while it cannot be comparable to that of covalent or ionic bonds. This intermediate strength is responsible for a wide variety of importance in chemistry and biology. For instance, H bond is commonly

T. Kumagai, *Visualization of Hydrogen-Bond Dynamics*, Springer Theses,
DOI: 10.1007/978-4-431-54156-1_1,

used in protein folding and the structures of DNA. H bond is also the path of H-atom transfer that is one of the most elementary and significant reactions in synthetic, environmental and biological processes. For instance, H-atom transfer is essential in biological catalysis because the transportation of H atom into the correct site and timing can enhance the process in enzyme reactions.

H bond in water is a quite ubiquitous question in nature. The H bond is formed between an H atom and an electronegative O atom, which provides unique properties of water [1], e.g., the relatively high boiling point, the lower density in solid than that in liquid, and the tendency to form dome-like droplets on solid surfaces. A water molecule can form four H bonds, thus two donating and two accepting H atoms. Although water molecules build three-dimensional networks in the liquid, continuous annihilation/re-creation of H bond and changes of molecular orientations/distances takes place with the time scale from femtosecond to picosecond. This flexibility of H bond gives rise to an abundant phase behavior in the pressure–temperature diagram. The dynamical fluctuation of H bond is also associated with various chemical and physical processes, e.g., solvation, acid–base reactions in solution, and freezing processes. For this reason, it has been a long-standing challenge to elucidate the net structure and dynamics of H bond in liquid water and numerous experimental [2–21] and theoretical [22–34] efforts have been devoted in the past. Infrared absorption, Raman scattering, depolarized light scattering, inelastic neutron scattering, and x-ray absorption spectroscopy have been employed to probe H-bond dynamics indirectly. The vibrational spectroscopies have been proven to be a powerful tool observing H-bonding nature and dynamics of water. A water molecule has three fundamental vibrations, i.e., the symmetric and asymmetric stretching modes at 3657 and 3756 cm^{-1}, respectively and the bending mode at 1595 cm^{-1} [34]. Especially, it is beneficial to detect the O–H stretching mode of water molecules because its frequency is quite sensitive to the H-bonding nature, like the number and relative strengths of H bond. Specifically, the frequency shows redshift as a water molecule forms H bond due to a weakening of the covalent OH bond. This weakening results from the substantial charge transfer from covalent O–H bond to the vicinity of H bond. In addition to the redshift, a spectral broadening can also arise from several reasons; anharmonic coupling to low-frequency modes, Fermi resonances with overtone and inhomogeneous broadening due to different H-bonding geometries [35–38]. Although linear vibrational spectroscopy provides a direct evidence of H-bonding interaction of steady states, it is quite difficult to gain insights of its dynamics because such spectroscopy gives only time-averaged signals. Ultrafast time-resolved vibrational spectroscopy is a powerful tool to investigate the dynamics of liquid water, which enables us to probe H-bond dynamics in real-time. Infrared pump-probe spectroscopy has often used to observe the ultrafast dynamics of the H-bond network of water [39–46]. However, the experimental data have often interpreted in only qualitative way. The difficulty to reproduce H-bonding dynamics mainly results from the complex potential energy landscape as well as a large number of possible network configurations. Moreover, quantum effects, i.e., tunneling, zero-point energy, become pronounced due to its small mass of H-atom, making it further complicated to clarify them. Experimental observations and theoretical descriptions of the H-bond dynamics including quantum effects have been attempted in homogeneous systems of gas, liquid, and condensed matter and consistent interpretations between experiment and theory has been established to some extent in the past decades. Especially, a water dimer in gas phase has been extensively studied as the model system because of its simple structure and most of the properties were determined well [47–50]; detailed in Chap. 5]. However H-bond dynamics in heterogeneous systems, like interface, has been unexplored so far, which is potentially related to the process of catalysis, electrode reaction, and energy productions.

1.2 Previous Studies of Water Adsorption on Metal Surfaces

There are numerous studies about water on surfaces. I briefly take a look back over studies of water adsorption on metal surfaces in past decades. Here I focus on studies carried out under ultra-high vacuum (UHV) conditions. The studies in well-defined environments have provided fundamental knowledge as to water adsorption on metal surfaces. The adsorption of water molecule on metal surfaces is related to some practical problems, e.g., corrosion, heterogeneous catalysis, hydrogen production/storage, and biological sensor. In addition, water-metal interfaces have been considered as a useful model system to understand the effect of truncating the three-dimensional H bond structure at the molecular level. Due to the symmetry similarity to the (0001) plane of ice I_h phase, single-crystalline closed-packed metal surfaces have been often employed to examine the growth and structure of ice film. The formation of H-bonded two-dimensional ice films, similar in structure to bulk ice, has been established in water adsorption on transition metal surfaces such as Pt, Ru, Ni, Pd and Rh.

The literature describing water adsorption on metal surfaces has been summarized in several reviews [51–54]. The early stage studies under UHV conditions were compiled by Thiel and Madey in 1987 [51] and renovated by Henderson in 2002 [52], where the structure of water layer and the interaction of water molecule with metal surfaces were main subjects. They summarized the so-called "bilayer model" of water adsorptions at metal surfaces (Fig. 1.1), which had been unmodified over the decades. In this model, water molecules form a network of interlinked hexamers where each molecule possesses three H bonds to its neighbors. Half of the water molecule donates both H atoms to the other molecules, whereas remains have either OH pointing towards surface (Fig. 1.1a) or vacuum (Fig. 1.1b). The direction of this remaining OH directly related to the growth mechanism of ice over metal surfaces. The ordered two-dimensional (2-D) structures were mainly investigated in earlier experiments using low-energy electron diffractions (LEED) [55–57]. Held and Menzel reported the first complete LEED-IV analysis of an ordered water layer on a Ru(0001) surface [58]. Ultraviolet photoelectron spectroscopy (UPS) and vibrational spectroscopy, i.e., electron energy loss spectroscopy (EELS) and infrared reflection absorption spectroscopy (IRAS), has been employed as very sensitive tools to characterize water molecules and its fragments and to determine the structure of water clusters and films on metal surfaces [59]. In EELS measurement for water adsorption on a platinum surface, three different O–H stretching modes at 2850 , 3380 and 3670 cm^{-1} at the sub-monolayer region was observed, corresponding to H-bonding towards the surface, between molecules and free O–H, respectively. Jacobi et al. investigated the development of water networks from monomers to three-dimensional ice using EELS with the improved vibrational resolution [60]. They showed H_2O molecules in the bilayer can be distinguished by two perpendicular vibrational modes that are observed 266 and 133 cm^{-1} for the upper and lower molecules, respectively. On the other hand, Sexton reported the complete isotopic exchange between H_2O and D_2O [61] on a platinum surface at 100 K, indicating a very high mobility of the H and D atoms within the water layer. Nagasaka et al. investigated the mobility of H in a two-dimensional H-bonding network consisting of H_2O and OH [62] using micro-scale x-ray photoelectron spectroscopy. They determined two H-transfer pathways with different time scales of 5.2 ± 0.9 ns and 48 ± 12 ns on a Pt(111) surface at 140 K. IRAS has been also employed to examine water adsorptions on metal surfaces. Griffiths et al. investigated H-bonded O-D stretching as a function of water coverage on three different metal surfaces, e.g., Ni(110), Pt(100) and Al(100) [63]. They found monomeric adsorption of D_2O molecules only on Ni(110) at 130 K, whereas clustering takes place very readily at low temperatures (130 K) and low covcrages on Pt(100). On Al(100), a broad distribution of cluster sizes was observed, which was attributed to the tendency of the limited mobility of D_2O molecules. More recently, Yamamoto et al. investigated monomeric adsorption and two- and three-dimensional islands growth of D_2O on a Rh(111) surface by means of IRAS [64]. The high resolution and sensitivity of IRAS made it possible to distinguish different clusters. They found that D_2O molecules form relatively small islands compared to Pt(111) and Ni(111) surfaces. This was attributed to the limited diffusion of water molecules on a Rh(111) surface due to the strong interaction between the water and the surface. Ogasawara et al. investigated the water adsorption on a Pt(111) surface at 100–120 K by means of X-ray absorption, emission and photoelectron spectroscopy [65]. They concluded that water molecules adsorb intact and form flat ice layer where the water molecules interact with the surface through the metal–oxygen (M–O) and metal–hydrogen (M–HO) bonds, like Fig. 1.1a.

For now it is commonly known that the structure of the first water layer is determined by a subtle balance between water–water interaction (H bond) and water-metal interactions. Specifically, the situation is divided into the following cases; (1) when the water–water interaction is stronger than water-metal one, water molecules form no wetting layer on the surface and three-dimensional growth takes place. (2) when their interactions are comparable, water molecules show abundant 2-D wetting layers depending on the registry of a surface. (3) when the water-metal interaction is much stronger, water molecules are dissociated on a surface. In fact, due to the competition of the interactions a wide variety of structures have been observed and characterized, like 2-D ice-like overlayers [66–69] 1-D chains

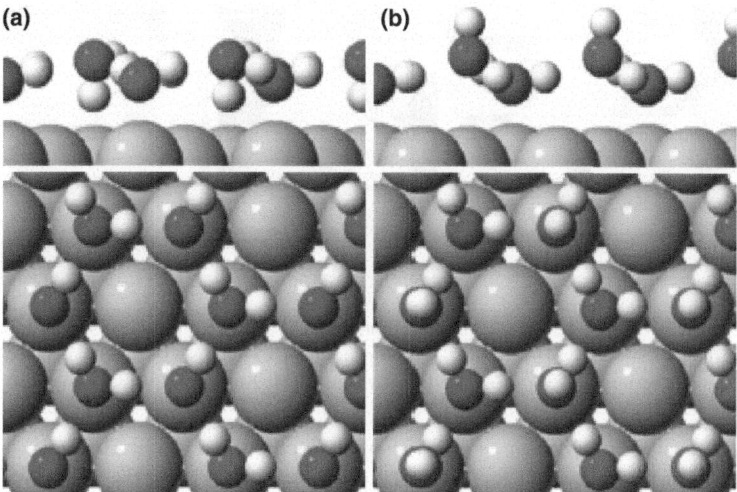

Fig. 1.1 Schematic drawing of water bilayer models adsorbed on a close packed metal surface. **a** An H-down geometry; One of H-atom in the upper half of the bilayer points down towards the metal surface. **b** An H-up geometry; One of H-atom in the upper half of the bilayer points into the vacuum. **c** Partially dissociated structure containing equal quantities of OH and water. In this case the O atoms are nearly co-planar and any H formed must either be adsorbed at the bare atop site or segregate to form separate domains. Reprinted with permission from Ref. [55]. Copyright 2009, Elsevier

[70–72], and mixed H_2O-OH structures [73–79]. The adsorption temperature is the other issue. At low temperatures there is insufficient thermal energy to allow water to diffuse on surfaces and water monomers and small clusters would be mainly formed. However, these structures are only kinetically stable and larger clusters are formed as the surface temperature is increased.

In more recent review, published in 2009 by Hodgson and Haq [54], the utility of STM to characterize small water clusters was highlighted. Morgenstern demonstrated imaging water molecules on metal surfaces by means of STM. In a sequence of papers, she showed the observation and characterization of individual water monomer and small clusters on Ag(111) [80–83], Cu(111) [84] surfaces. It was found that the H_2O molecules simultaneously form an apparently zero-dimensional structure consisting of cyclic hexamer and H-bonded networks with different dimensionalities on a Ag(111) surface. Furthermore, STM was employed to probe the dynamical property of individual small water clusters [85]. Mitsui et al. investigated the adsorption, clustering and diffusion properties of water molecules on Pd(111) using a variable-temperature STM. They found water molecules adsorb mainly as a monomer at low temperature (40 K) and coverages and diffuse on the surface. The diffusion caused the collision between monomers and clustering. Interestingly, the diffusion rate was found to be very different by several orders of magnitude depending on the cluster size. Being common to both Ag(111) and Pd(111) surfaces, cyclic hexamers were found to be particularly

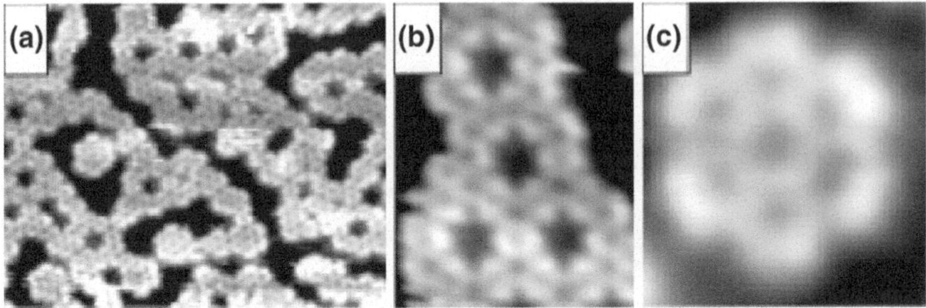

Fig. 1.2 **a** STM image of 2-D ice film formed on Pd(111) at 100 K (175 × 135 Å^2). **b** Zoom-in image of the lace structure (41 × 53 Å^2). **c** Zoom-in image of the rosette structure (20 × 20 Å^2). Reprinted with permission from Ref. [69]. Copyright 2004, American Physical Society

stable. The characterization of individual molecules and small clusters is extremely difficult by means of conventional spectroscopy. STM opens up a novel way to investigate the structure and dynamics of water molecules on metal surfaces at the single molecule level. STM studies also provided visible insights to 2-D wetting layers [69]. Figures 1.2 show the STM images of ice layer formed on Pd(111) at 100 K where the ice-like hexagonal pattern is clearly resolved. Along with the development of the STM techniques, large-scale computations have been also developed to examine the adsorption structure of water in the last decade [86–92]. Since it is quite difficult to determine the molecular orientations solely by STM experiments, the calculations based on the density functional theory have been generally employed to quantitatively interpret STM images. Both STM experiments and theoretical calculations found that a water molecule adsorbs at a top site on metal surfaces. Calculations also predicted the water monomer binding energy increases in order of Au $<$ Ag $<$ Cu $<$ Pd $<$ Pt $<$ Ru $<$ Rh.

1.3 Single Atom/Molecule Science

Characterization of single atoms and molecules has been a scientifically attractive and challenging subject. Probing individual atoms and molecules makes it possible to unveil fundamental properties which are hidden behind an ensemble of atoms and molecules. The important transition of the field of single atom/molecule science was provided by the invention of STM [93] and its derivatives, namely scanning probe microscope (SPM).

Feynman pointed out "There's plenty of room at the Bottom" in his talk on 29th, December in 1959 at the annual meeting of the American Physical Society at California Institute of Technology [94], which is often regarded as the begging of nanotechnology. He believed in *plenty of room* for the findings of new phenomena and the development of bland-new devices working at the atomic scale. Over the last few decades the research field of *nanoscience* and *nanotechnology* has attracted a great deal of attentions and we have witnessed massive evolutions of these fields [95]. Feynman also mentioned a sorting of individual atoms in his talk. This "atom manipulation" was achieved in 1989 by Eigler and Schweitzer at IBM Almaden Research Center. They demonstrated positioning individual atoms at the atomic-scale precision using the low temperature STM and succeeded in writing *atomic letters*, "I-B-M", on a metal surface with 35 individual xenon atoms [96]. In their report xenon atoms were transferred back and forth reversibly between the STM tip and the surface. They also found the electrical conductance of the STM junction switches between low and high states depending on the xenon position (on the tip or the surface), which can be considered as an atomic scale switch. Furthermore, the atom manipulation was applied assembling artificial nano-scale structure to observe electronic [97, 98] and spin [99–101] states at the scale.

The studies of surface-mediated reactions at the single molecule level will contribute to establish the principles of catalytic phenomena. STM has been employed to investigate the single molecule reactions. Wintterlin, Schuster, and Ertl at Fritz-Haber Institute reported the dissociation of single oxygen molecules on a platinum surface using variable temperature STM in 1996 [102]. They revealed the temperature dependence of the dissociation and diffusion dynamics of molecular and atomic oxygen. In the late 1990s and early 2000s, Ho and his co-workers opened a novel field of *single molecule chemistry* [103]. They demonstrated the capability of STM to control reactions of single molecules on surfaces and to identify the chemical species. The dissociation of oxygen molecules on platinum surface was induced by injection of low-energy electrons (0.2–0.4 V) from STM tip [104]. The chemical sensitivity of STM was ameliorated by combining with inelastic electron tunneling spectroscopy (IETS) that is a sort of vibrational spectroscopy [105]. They also demonstrated the formation and characterization of single bond [106]. The bond formation between a single iron atom and carbon monoxide molecule was controlled using STM manipulation and its bonding nature was characterized by STM-IETS. A multi-step reaction of single molecules was also demonstrated by Hla et al. [107]. They controlled the selective C-I bond dissociation of iodobenzene molecules and the subsequent dimerization yielding a biphenyl using STM tip. These pioneering efforts indicated the possibility to control chemistry at the spatial limit of individual atoms and molecules.

Although it is out of scope of this thesis, recent progress in optical microscopy has also enabled us to detect single molecules, which forms another research frontier. Optical methods possess a capability of high temporal resolution that is suitable for investigating dynamical behavior, but the spatial resolution is less than SPM. The detection of single molecules is conducted in very dilute samples with conventional optical microscopies and spectroscopies.

The potential application of single atom/molecule science would be molecular electronics that are a rapidly-growing field [108]. One of the key questions is if it is possible to use single molecules as active elements in nano-scale circuits for many applications. SPM-based techniques will continue to apply for examining such devices.

1.4 The Motivation and Scope of this Thesis

The motivation of this thesis is to exploit H-bond dynamics in terms of the single molecule level using STM. STM capabilities, i.e., imaging, manipulation, chemical modification and spectroscopic characterization for individual molecules, are applied to observe, assemble and characterize water-based model systems on a metal surface in my study. The scope of this thesis is imaging and characterization of a single water molecule, small clusters, dissociated products (hydroxyl groups) and water-hydroxyl complexes on a metal surface. The basic principles and techniques of STM are described in Chap. 2. The experimental setup and sample preparation is detailed in Chap. 3. The results and discussions are described in Chap. 4–10. Finally I summarize the conclusions in Chap. 11.

References

1. V.F. Petrenko, R.W. Whitworth, *Physics of ice, Oxford* (Oxford Univ Press, Oxford, 1998)
2. C.J. Montrose, J.A. Bucaro, J. Marshall-Coakley, T.A. Litovitz, J. Chem. Phys. **60**, 5025 (1974)
3. W. Danninger, G. Zundel, J. Chem. Phys. **74**, 2769 (1981)
4. S.H. Chen, J. Teixeira, R. Nicklow, Phys. Rev. A **26**, 3477 (1982)
5. J. Teixeira, M.C. Bellissent-Funel, S.H. Chen, A.J. Dianoux, Phys. Rev. A **31**, 1913 (1985)
6. J. Teixeira, Mol. Phys. **110**, 249 (2012)
7. R. Jimenez, G.R. Fleming, P.V. Kumar, M. Maroncelli, Nature **369**, 471 (1994)
8. S. Woutersen, U. Emmerichs, H.J. Bakker, Science **278**, 658 (1997)
9. S. Woutersen, H.J. Bakker, Nature **402**, 507 (1999)
10. M.F. Kropman, H.J. Bakker, Science **291**, 2118 (2001)
11. Y.L.A. Rezus, H.J. Bakker, J. Chem. Phys. **123**, 114502 (2005)
12. G.E. Walrafen, J. Phys. Chem. **94**, 2237 (1990)
13. E.W. Castner, Y.J. Chang, Y.C. Chu, G.E. Walrafen, J. Chem. Phys. **102**, 653 (1995)
14. C.P Lawrence, J.L Skinner, Chem. Phys. Lett. 369, 472 (2003).
15. A. Pakoulev, Z. Wang, Y. Pang, D.D. Dlott, Chem. Phys. Lett. **378**, 281 (2003)
16. A. Pakoulev, Z. Wang, Y. Pang, D.D. Dlott, Chem. Phys. Lett. **380**, 404 (2003)
17. J.J. Loparo, S.T. Roberts, A. Tokmakoff, J. Chem. Phys. **125**, 194522 (2006)
18. J. Stenger, D. Madsen, P. Hamm, E.T. Nibbering, T. Elsaesser, Phys. Rev. Lett. **87**, 027401 (2001)
19. E.T.J. Nibbering, T. Elsaesser, Chem. Rev. **104**, 1887 (2004)
20. D. Kraemer, M.L. Cowan, A. Paarmann, N. Huse, E.T.J. Nibbering, T. Elsaesser, R.J.D. Miller, Proc. Natl. Acad. Sci. **105**, 437 (2008)
21. N. Huse, H. Wen, D. Nordlund, E. Szilagyi, D. Daranciang, T.A. Miller, A. Nilsson, R.W. Schoenlein, A.M. Lindenberg, Phys. Chem. Chem. Phys. **11**, 3951 (2009)
22. D. Eisenberg, W. Kauzmann, *The Structure and Properties of Water* (Oxford Univ. Press, New York, 1969)
23. V. Mazzacurati, M.A. Ricci, G. Ruocco, M. Sampoli, Chem. Phys. Lett. **159**, 383 (1989)
24. A. Luzar, D. Chandler, Phys. Rev. Lett. **76**, 928 (1996)
25. C.H. Cho, S. Singh, G.W. Robinson, Faraday Discuss. **103**, 19 (1996)
26. I. Omine, H. Tanaka, Chem. Rev. **93**, 2545 (1993)
27. I. Ohmine, S. Saito, Acc. Chem. Res. **32**, 741 (1999)
28. I. Ohmine, J. Phys. Chem. **99**, 6767 (1995)

29. M. Matsumoto, S. Saito, I. Ohmine. Nature **416**, 409 (2002)
30. D. Marx, M.E. Tuckerman, J. Hutter, M. Parrinello, Nature **397**, 601 (1999)
31. M.E. Tuckerman, D. Marx, M. Parrinello, Nature **417**, 925 (2002)
32. M.E. Tuckerman, D. Marx, M.L. Klein, M. Parrinello, Science **275**, 817 (1997)
33. R. Rey, K.B. Møller, J.T. Hynes, J. Phys. Chem. A **106**, 11993 (2002)
34. G. Herzberg, *Molecular Spectra and Molecular Structure* (van Nostrand, New York, 1945)
35. Y. Maréchal, A.J. Witkowski, J. Chem. Phys. **48**, 3697 (1968)
36. Y. Maréchal, J. Chem. Phys. **87**, 6344 (1987)
37. D. Chamma, O. Henri-Rousseau, Chem. Phys. **248**, 53 (1999)
38. O. Henri-Rousseau, P. Blaise, D. Chamma, Adv. Chem. Phys. **121**, 241 (2002)
39. G.M. Gale, G. Gallot, F. Hache, N. Lascoux, S. Bratos, J.-C. Leicknam, Phys. Rev. Lett. **82**, 1068 (1999)
40. G. Gallot, S. Bratos, S. Pommeret, N. Lascoux, J.C. Leicknam, M. Kozinski, W. Amir, G.M. Gale, J. Chem. Phys. **117**, 11301 (2002)
41. C.J. Fecko, J.D. Eaves, J.J. Loparo, A. Tokmakoff, P.L. Geissler, Science **301**, 1698 (2003)
42. R. Laenen, C. Rauscher, A. Laubereau, Phys. Rev. Lett. **80**, 2622 (1998)
43. R. Laenen, C. Rauscher, A. Laubereau, J. Phys. Chem. B **102**, 9304 (1998)
44. D. Cringus, S. Yeremenko, M.S. Pshenichnikov, D.A. Wiersma, J. Phys. Chem. B **108**, 10376 (2004)
45. J.C. Deak, S.T. Rhea, L.K. Iwaki, D.D. Dlott, J. Phys. Chem. A **104**, 4866 (2000)
46. A.J. Lock, H.J. Bakker, J. Chem. Phys. **117**, 1708 (2002)
47. N. Pugliano, R.J. Saykally, J. Chem. Phys. **96**, 1832 (1992)
48. N. Pugliano, R.J. Saykally, Science **257**, 1937 (1992)
49. J.D. Cruzan, et al., ibid. 271, 59 (1996); K. Liu, M.G. Brown, J.D. Cruzan, R.J. Saykally, ibid., p. 62
50. J. K. Gregory, D. C. Clary, K. Liu, M. G. Brown and R. J. Saykally, ibid, 275, 5301 (1997).
51. P.A. Thiel, T.E. Madey, Surf. Sci. Rep. **7**, 211 (1987)
52. M.A. Henderson, Surf. Sci. Rep. **46**, 1 (2002)
53. A. Verdaguer, G.M. Sacha, H. Bluhm, M. Salmeron, Chem. Rev. **106**, 1478 (2006)
54. A. Hodgson, S. Haq, Surf. Sci. Rep. **64**, 381 (2009)
55. L.E. Firment, G.A. Somorjai, J. Chem. Phys. **63**, 1037 (1975)
56. L.E. Firment, G.A. Somorjai, Surf. Sci. **84**, 275 (1979)
57. D.L. Doering, T.E. Madey, Surf. Sci. **123**, 305 (1982)
58. G. Held, D. Menzel, Surf. Sci. **316**, 92 (1994)
59. H. Ibach, S. Lehwald, Surf. Sci. **91**, 187 (1980)
60. K. Jocobi, K. Bedürftig, Y. Wang, G. Ertl, Surf. Sci. **472**, 9 (2001)
61. B.A. Sexton, Surf. Sci. **94**, 435 (1980)
62. M. Nagasaka, H. Kondoh, K. Amemiya, T. Ohta, Y. Iwasawa, Phys. Rev. Lett. **100**, 106101 (2008)
63. K. Griffiths, R.V. Kasza, F.J. Esposto, B.W. Callen, S.J. Bushby, P.R. Norton, Surf. Sci. **307–309**, 60 (1994)
64. S. Yamamoto, A. Beniya, K. Mukai, Y. Yamashita, J. Yoshinobu, J. Chem. Phys. B **109**, 5816 (2005)
65. H. Ogasawara, B. Brena, D. Nordlund, M. Nyberg, A. Pelmenschikov, L.G.M. Pettersson, A. Nilsson, Phys. Rev. Lett. **89**, 276102 (2002)
66. T. Schiros, S. Haq, H. Ogasawara, O. Takahashi, H. Östrom, K. Andersson, L.G.M. Pettersson, A. Hodgson, A. Nilsson, Chem. Phys. Lett. **429**, 415 (2006)
67. G. Zimbitas, S. Haq, A. Hodgson, J. Chem. Phys. **123**, 174701 (2005)
68. C. Clay, S. Haq, A. Hodgson, Chem. Phys. Lett. **388**, 39 (2004)
69. J. Cerdá, A. Michaelides, M.L. Bocquet, P.J. Feibelman, T. Mitsui, M. Rose, E. Fomin, M. Salmeron, Phys. Rev. Lett. **93**, 116101 (2004)
70. T. Yamada, S. Tamamori, H. Okuyama, T. Aruga, Phys. Rev. Lett. **96**, 036105 (2006)
71. J. Carrasco, A. Michaelides, M. Forster, S. Haq, R. Raval, A. Hodgson, Nature Mat. **8**, 427 (2009)

72. M. Morgenstern, T. Michely, G. Comsa, Phys. Rev. Lett. **77**, 703 (1996)
73. J. Lee, D.C. Sorescu, K.D. Jordan, J.T. Yates Jr, J. Chem Phys, C **112**, 17672 (2008)
74. M. Forster, R. Raval, A. Hodgson, J. Carrasco, A. Michaelides, Phys. Rev. Lett. **106**, 046103 (2011)
75. C. Clay, A. Hodgson, Curr. Opin. Solid State Mater. Sci. **9**, 11 (2005)
76. C. Clay, S. Haq, A. Hodgson, Phys. Rev. Lett. **92**, 046102 (2004)
77. J. Weissenrieder, A. Mikkelsen, J.N. Andersen, P.J. Feibelman, G. Held, Phys. Rev. Lett. **93**, 196102 (2004)
78. G. Held, C. Clay, S.D. Barrett, S. Haq, A. Hodgson, J. Chem. Phys. **123**, 064711 (2005)
79. L. Guillemot, K. Bobrov, J. Phys. Chem. C **115**, 22387 (2011)
80. K. Morgenstern, Surf. Sci. **504**, 293 (2002)
81. K. Morgenstern, J. Nieminen, Phys. Rev. Lett. **88**, 066102 (2002)
82. K. Morgenstern, J. Nieminen, J. Chem. Phys. **120**, 10786 (2004)
83. H. Gawronski, J. Carrasco, A. Michaelides, K. Morgenstern, Phys. Rev. Lett. **101**, 136102 (2008)
84. K. Morgenstern and K-H. Rieder, J. Chem. Phys. 116, 5746 (2002).
85. T. Mitsui, M.K. Rose, E. Fomin, D.F. Ogletree, M. Salmeron, Science **297**, 1850 (2002)
86. A. Michaelides, P. Hu, J. Am. Chem. Soc. **123**, 4235 (2001)
87. A. Michaelides, P. Hu, J. Chem. Phys. **114**, 513 (2001)
88. A. Michaelides, A. Alavi, D.A. King, J. Am. Chem. Soc. **125**, 2746 (2003)
89. P.J. Feibelman, Science **295**, 99 (2002)
90. G.S. Karlberg, F.E. Olsson, M. Persson, G. Wahnstrom, J. Chem. Phys. **119**, 4865 (2003)
91. A. Michaelides, Appl. Phys. A **85**, 415 (2006)
92. A. Michaelides, V.A. Ranea, P.L. de Andres, D.A. King, Phys. Rev. Lett. **90**, 216102 (2003)
93. G. Binnig, H. Rohler, Helv. Phys. Acta **55**, 725 (1982)
94. http://www.zyvex.com/nanotech/feynman.htlm.
95. C. Joachim and L. Plévert, *Nanoscience: The invisible revolution* (World Scientific, 2008)
96. D.M. Eigler, E.K. Schweizer, Nature **344**, 524–526 (1990)
97. M.F. Crommie, C.P. Lutz, D.M. Eigler, Science **262**, 218 (1993)
98. M.F. Crommie, C.P. Lutz, D.M. Eigler, Nature **363**, 524 (1993)
99. C.F. Hirjibehedin, C.P. Lutz, A.J. Heinrich, Science **312**, 1021 (2006)
100. S. Loth, M. Etzkorn, C.P. Lutz, D.M. Eigler, A.J. Heinrich, Science **329**, 1628 (2010)
101. S. Loth, S. Baumann, C.P. Lutz, D.M. Eigler, A.J. Heinrich, Science **335**, 196 (2012)
102. J. Wintterlin, R. Schuster, G. Ertl, Phys. Rev. Lett. **77**, 123 (1996)
103. W. Ho, J. Chem. Phys. **117**, 11033 (2002)
104. B.C. Stipe, M.A. Rezaei, W. Ho, S. Gao, M. Persson, B.I. Lundqvist, Phys. Rev. Lett. **78**, 4410 (1997)
105. B.C. Stipe, M.A. Rezaei, W. Ho, Science **280**, 1732 (1998)
106. H.J. Lee, W. Ho, Science **286**, 1719 (1999)
107. S.-W. Hla, L. Bartels, G. Meyer, K.-H. Rieder, Phys. Rev. Lett. **85**, 2777 (2000)
108. J.C. Cuevas, E. Scheer, *Molecular electronics—An introduction to theory and experiment* (World Scientific Publishing Co. Pte, Ltd, 2010)

Chapter 2
Principles and Techniques

Abstract I describe a brief summary of quantum mechanics and principles of scanning tunneling microscopy (STM) in this chapter. Quantum tunneling is an important concept to describe the transportation of small particles, like electron and hydrogen. In Sect. 2.1 I focus on H-atom tunneling in a double minimum potential that is the simplest but a ubiquitous system in physics and chemistry. Quantum tunneling of electron directly relates to the principle of the STM. In Sect. 2.2 the principles and applications of STM are described. I focus on the inelastic electron tunneling process in an STM junction, which can be applied to the vibration spectroscopy and reaction control of single molecules.

Keywords Quantum tunneling · Double minimum potential · Scanning tunneling microscope/spectroscopy · Inelastic electron tunneling

2.1 Quantum Mechanics

2.1.1 Basics of Quantum Tunneling

Quantum tunneling is derived from the concept of quantum mechanics, where a particle can pass through a barrier that is classically insurmountable. This is a result of the wave-particle duality of matter. Quantum tunneling becomes pronounced in light particles such as electron and H atom.

Figure 2.1 shows the simplest model in which a particle tunnels through a one-dimensional rectangular barrier. The time-independent Schrödinger equation of the particle is written by

$$-\frac{\hbar^2}{2m}\frac{d^2}{dx^2}\psi(x) + V(x)\psi(x) = E\psi(x) \tag{2.1}$$

T. Kumagai, *Visualization of Hydrogen-Bond Dynamics*, Springer Theses,
DOI: 10.1007/978-4-431-54156-1_2,

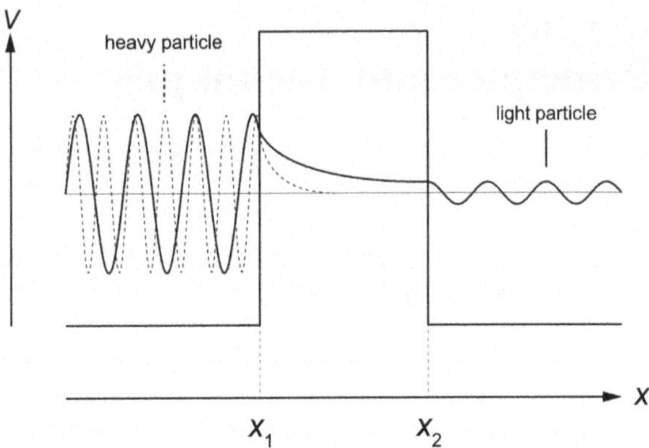

Fig. 2.1 Schematic diagram of one-dimensional rectangular potential barrier

$$\frac{d^2}{dx^2}\psi(x) = \frac{\hbar^2}{2m}(V(x) - E)\psi(x) \tag{2.2}$$

where m is the mass, $\psi(x)$ is the wave function, E is the energy of the particle and $V(x)$ is the potential. In the region of $V(x) < E$

$$\psi(x) = Ae^{-kx} + Be^{kx} \tag{2.3}$$

$$k^2 = -\frac{2m(V(x) - E)}{\hbar^2} \tag{2.4}$$

The wave function shows the oscillation characterized by the wave vector k. On the other hand, in the region of $V(x) > E$

$$\psi(x) = Ae^{-\kappa x} + Be^{\kappa x} \tag{2.5}$$

$$\kappa^2 = \frac{2m(V(x) - E)}{\hbar^2} \tag{2.6}$$

The wave function no longer shows the oscillation, and it monotonically decreases with the decay constant κ. For light particles we have a chance to find the particle beyond the classically impenetrable barrier, and the wave function starts oscillating again. The transmission probability T of a particle can be analytically-derived in this simple model, which is given

$$T = \left\{1 + \frac{(e^{\kappa L} - e^{-\kappa L})^2}{16\varepsilon(1 - \varepsilon)}\right\}^{-1} \tag{2.7}$$

where $\varepsilon = E/V$, and $L = (x_2\text{-}x_1)$ is the barrier width of a one-dimensional rectangular potential. If we assume $\kappa L \gg 1$, corresponding to a barrier of sufficient height and width,

$$T \approx 16\varepsilon(1 - \varepsilon)e^{-2\kappa L} \tag{2.8}$$

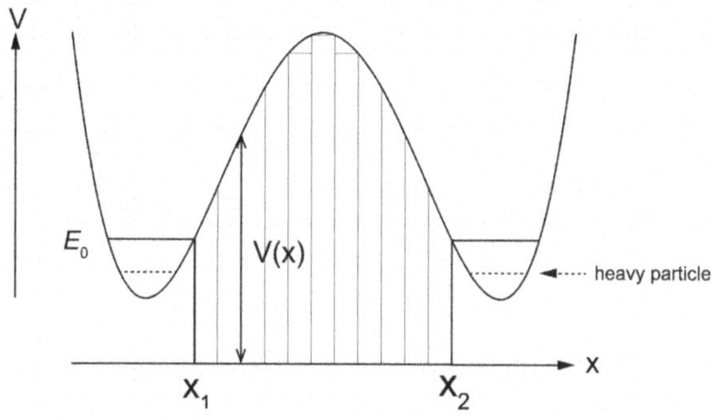

Fig. 2.2 Schematic diagram of one-dimensional smooth potential barrier

The transmission probability decreases exponentially with an increase of the mass, $m^{1/2}$, and the barrier width, *L*. Thus a significant isotope effect between H and D (deuterium) is expected in the dynamical processes.

The one-dimensional rectangular barrier is an oversimplified model system. The barrier that smoothly varies, as shown in Fig. 2.2, is a more realistic question. Using the WKB approximation, the tunneling probability *D*(*E*) of a particle that has the total energy *E* in a one-dimensional potential barrier *V*(*x*) is expressed by

$$D(E) = \exp\left\{-\frac{2}{\hbar}\int_{x_1}^{x_2}\sqrt{2m(V(x)-E)}dx\right\} \tag{2.9}$$

where x_1 and x_2 is the classical turning points, and (x_1-x_2) corresponds to the barrier length. In the WKB approximation the Planck constant is regarded as a small number. Therefore this approximation is called semi-classical and valid when the total energy is significantly lower than the barrier. It is noted that the WKB approximation is restricted to one-dimensional problems, thus it cannot be applied to multi-dimensional problems that are much more likely in chemical and biological processes.

In addition to the mass effect, the zero point energy (ZPE) must be considered when we treat the kinetics of H and D. Assuming a harmonic potential, the ZPE is given by

$$E_0 = \frac{1}{2}\hbar\omega, \omega = \sqrt{\frac{k}{m}} \tag{2.10}$$

where ω is the vibrational frequency and *k* is the spring constant. According to (2.10), the ZPE of D atom is smaller by the factor of $(1/2)^{1/2}$ than that of H. This reduction of the ZPE also lowers the transmission probability of D atom through the increase of the effective barrier.

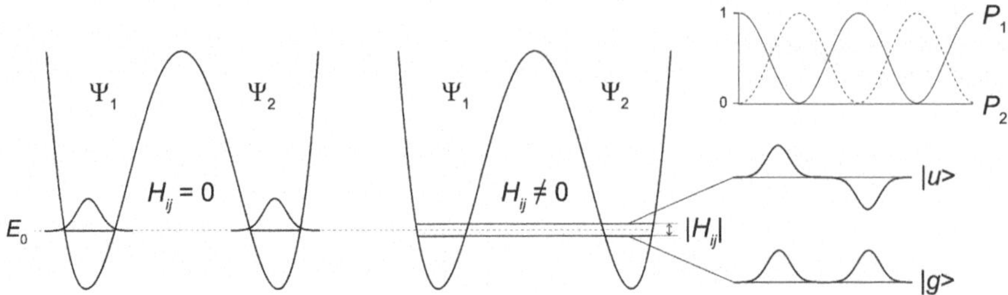

Fig. 2.3 Schematic illustration of a symmetric double minimum potential. **a** $H_{\mathrm{ij}} = 0$ **b** $H_{\mathrm{ij}} \neq 0$

2.1.2 Tunneling in a Symmetric Double Minimum Potential

Symmetry is important to understand diverse concepts and laws of nature. Here I focus on the role of symmetry in H transfer. The symmetry of the potential landscape is a key to the mechanism involving quantum tunneling. The simplest, but a ubiquitous system in physics and chemistry is a symmetric double minimum potential (Fig. 2.3), where we can consider two wave functions $|\Psi_1>$ and $|\Psi_2>$ with a specific Hamiltonian H_i ($i = 1$ or 2) in each potential well where the eigenenergy is given by $H_i\Psi = E_0\Psi_i$. If each of the two states is independent, the wave packet is localized in each of the potential well for all time. Now we take into account the transfer of a wave packet between two wells and introduce the transition matrix $\hat{H}_{ij}(i, j = 1 \text{or } 2;\ i \neq j)$ that transfers the wave packet into another well. As long as the matrix element $H_{ij} = <\Psi_i|\hat{H}_{ij}|\Psi_j> = 0$, each of the two states degenerates into the eigenstate with its eigenenergy of E_0 (Fig. 2.3a). However, this situation is varied if $H_{ij} \neq 0$ (Fig. 2.3b), where $|\Psi_1>$ and $|\Psi_2>$ are no longer the eigenstate of the system and the eigenfunctions are given by the symmetric (gerade) and anti-symmetric (ungerade) linear combination of $|\Psi_1>$ and $|\Psi_2>$.

$$|g> = \frac{1}{\sqrt{2}}(|\Psi_1> + |\Psi_2>),\ |u> = \frac{1}{\sqrt{2}}(|\Psi_1> - |\Psi_2>)$$

The corresponding energy levels are split into

$$E_k = E_0 \pm |H_{12}|$$

where k is g (gerade) or u (ungerade) and the lower or upper energy levels belongs to the symmetric or anti-symmetric state. $|H_{12}|$ corresponds to the energy splitting. Given that the system is initially in the state of $|\Psi_1>$, the system shows a periodic motion between $|\Psi_1>$ and $|\Psi_2>$ states because the $|\Psi_1>$ is no longer an eigenstate of the system. Then the probabilities finding the wave packet in $|\Psi_1>$ and $|\Psi_2>$ change according to

$$P_1 = \cos^2(H_{12}|t),\ P_2 = \sin^2(H_{12}|t)$$

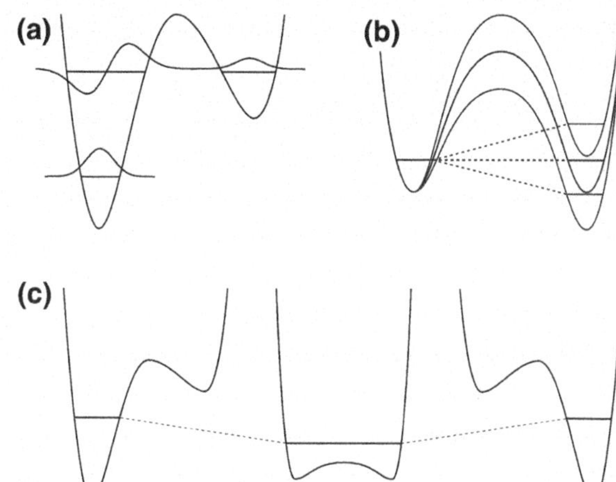

Fig. 2.4 Schematic illustration of asymmetric double minimum potentials. **a** In condensed matter intermolecular interactions tune a symmetric potential. **b** In liquid the potential is continuously perturbed through surrounding solvents. **c** An alternative scheme of H atom transfer in liquid. The strong interaction causes the significantly reduced barrier where the H atom is adiabatically transferred

The frequency of this oscillation v_t, the so-called tunneling frequency, is given by

$$v_t = \frac{|H_{12}|}{\pi}$$

This periodic oscillation is called the coherent tunneling.

Hydrogen (proton) tunneling has been derived from the theory of the one-dimensional symmetric double oscillator [1]. The separation between the two states are characterized by $\Delta E = hv_t$. (h is Planck's constant). The transfer time τ is related to the tunneling frequency by $\tau = 1/2v_t^{-1}$. The tunneling splitting strongly depends on the barrier height and width. The energy level splitting due to coherent tunneling has been observed in many model systems such as small water clusters [2], NH_3 [3], tropolone [4, 5], and malonaldehyde [6–8] in the gas phase, using rotation-vibration spectroscopy. The tunneling splitting, namely the tunneling frequency, varies from 10^{12} Hz to a few Hz depending on the system.

2.1.3 Tunneling in an Asymmetric Double Minimum Potential

Surrounding environments or external perturbations alter the symmetry of the potential landscape of H transfer. In condensed matter, intermolecular interactions commonly vary the symmetric potential, resulting in an asymmetric potential as depicted in Fig. 2.4a. In this situation, the wave packet of hydrogen is forced to be localized in one potential well. However, tunneling is still possible at the excited state. This situation is depicted in Fig. 2.4a. A vibrational excitation generates the wave packet in the excited state, which is further transferred to another well via tunneling. This process is no longer coherent.

In liquid, the situation becomes much more complicated. The effective symmetry of the potential is continuously altered by surrounding solvents. This situation is depicted in Fig. 2.4b. An alternative description about H atom transfer in liquid was proposed by Borgis and Hynes [9, 10]. As shown in Fig. 2.4c the asymmetric single or the double well potential, where the wave packet of hydrogen is localized in one well, is continuously converted into symmetric potential by surrounding solvents. But the strong interaction between the molecules results in a significantly reduced barrier, where the ZPE overcomes the barrier. In this situation H transfer takes place adiabatically and hydrogen motion is no longer governed by tunneling.

2.2 Scanning Tunneling Microscope

2.2.1 Overview

STM is a powerful tool to image conductive surfaces with an atomic resolution. In principle, an optical microscope cannot realize such a high resolution due to its diffraction limit. STM was developed in 1982 by Binnig and Rohrer at IBM Zürich Research Laboratory [11]. They were awarded the Nobel Prize in 1986 for the development of the STM. In STM, the highest resolution is considered to be 0.1 nm laterally and 0.01 nm vertically. STM set a new direction for nanoscience and nanotechnology and has continued to evolve as a fascinating tool to investigate physical, chemical, and biological processes at the spatial limit.

STM is based on the concept of quantum tunneling of electron. When a conducting tip having an atomically sharp apex, is brought very close (few Å) to the conductive surface while a bias voltage is applied between the tip and the surface, electrons can tunnel through the barrier between them. The tunneling current is described as a function of the relative distance between the tip and surface, the applied voltage, and the local density of states (LDOS) of the tip and surface. Since the tunneling current depends exponentially onto the tip-surface distance, the current has a considerably sensitivity to the surface corrugation. If the tip apex consists of a single or few atoms, it has an atomic level sensitivity against the surface corrugation. The STM image is acquired by scanning the surface with controlling the distance between the tip and surface. Figure 2.5 shows a schematic illustration of the basic STM setup. The STM measurement is carried out by following manner; first, bias voltage is applied between a sharp tip and a conductive substrate, and then the tip is brought close to the substrate using a coarse piezo control system of the tip along z direction (surface normal). This coarse motion is turned off when the tip and the surface become sufficiently close (few Å) and the tunneling current is detected. At the tunneling region, the fine control of the tip in all three dimensions is required to maintain the tip-sample distance d ($3 < d < 10$Å). Once the tunneling junction is established, the bias voltage and the tip position with respect to the surface can be varied. The usual bias voltage to image molecules chemisorbed on surfaces is in a range of ± 2 V. Two different modes to obtain STM image exist. First, the tip is scanned with constant z height (constant height mode).

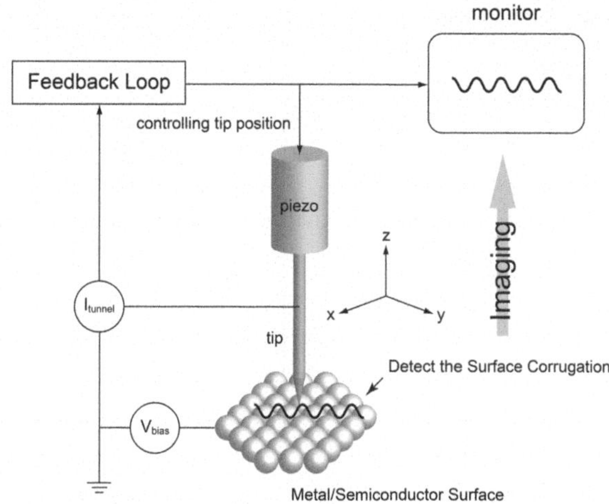

Fig. 2.5 Schematic illustration of a basic setup of the STM

The variation of the tunnel current is recorded as the tip is scanned along the surface. Second, the tip is scanned with the tunneling current kept constant during the scanning (constant current mode). In this case, the z position of the tip is controlled via the feedback loop and it varies during the measurements, and this variation of the z position is recorded to produce an STM image. Thus the STM image corresponds to topography across the surface and gives approximately constant LDOS of the surface; this means the contrast of the image is due to variations in the density of states. The constant height mode, with the voltage and height both kept constant, gives an image of current variation over the surface, which can also be related to the LDOS. The benefit of using a constant height mode is that it can record an image faster as the tip is not require to precisely follow the corrugation of the surface. Thus the constant height mode is usually employed in a high speed measurement which enables us to record an image at a video rate.

In addition to imaging a conductive surface, since the tunneling current stems from the LDOS, we can extract the local electronic structure of a surface by sweeping the bias voltage and measuring current with fixed the tip at a specific location. This measurement is called scanning tunneling spectroscopy (STS). The technical details and the applications of the STM were already summarized in several good books [12–15]. STM and its derivative techniques, called the scanning probe technique including the atomic force microscopy and the scanning near optical field microscopy, will keep on evolving as a base technology of nanoscience.

2.2.2 Tunneling Current

STM relies on the electric current originated from quantum tunneling of electron flowing between a conductive tip and surface. For a proper interpretation of an

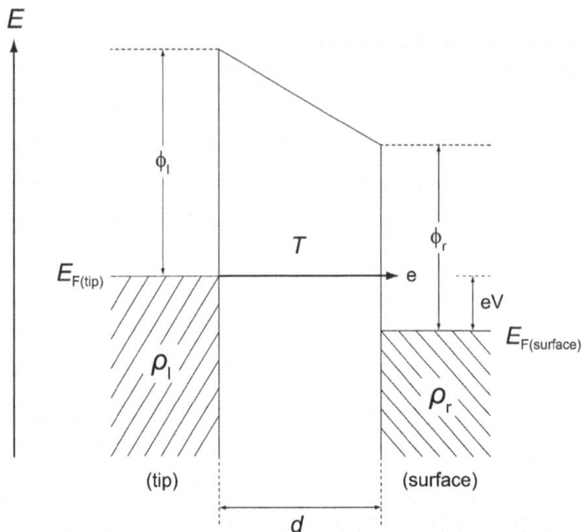

Fig. 2.6 Schematic illustration of tunnel junction in STM

STM image it is essential to understand how tunneling current is described in the STM junction. Figure 2.6 illustrates a typical tunnel junction consisted of metal–insulator (vacuum gap)-metal system. Tunneling current can be expressed by

$$I \propto \int_0^{eV} e\rho_l(E)\rho_r(-eV+E)T(R,eV,E)dE \tag{2.11}$$

$\rho_{l(r)}$ is the density of state of the left(right) electrodes, corresponding to a conductive substrate and tip in STM, *E* is the energy of electrodes with respect to the Fermi level, *V* is a bias voltage applied between the tip and surface, and *T* is a tunnel probability which can be expressed using the WKB approximation,

$$T = \exp\left\{-\frac{2d\sqrt{2m}}{\hbar}\sqrt{\frac{\phi_l+\phi_r}{2}-E+\frac{eV}{2}}\right\} \tag{2.12}$$

where *d* is the distance between the electrodes, thus the thickness of the potential barrier, $\phi_{l(r)}$ is the work function of the electrodes. From Eq. (2.12), tunneling current shows an exponential dependence onto the electrode distance. As a consequent, tunneling current is very sensitive to *d*.

According to Eq. (2.11) the tunneling current in STM reflects the LDOS of a tip and substrate. If the change in an LDOS of a substrate is induced by the adsorption of molecules, tunneling current is significantly affected in the vicinity of the molecule. With the adsorption of molecule, the charge transfer between the molecule and surface occurs, inducing molecule-induced state near the Fermi level. However, the interpretation of an STM image of molecules on a surface is not straightforward. For instance, carbon monoxide molecule is observed as depression in STM image, meaning the reduction of the density of state near the Fermi level. On the other hand, water molecule is observed as protrusion, meaning the increase of the DOS near the Fermi level. The explanation has been demonstrated by using theoretical calculations in the past [16]. The appearance of STM image also depends on the tip conditions. For instance, if a molecule adsorbed on the tip apex, the STM image is significantly changed. Therefore we have to discuss an STM image very carefully and a combination of first principle calculations is quite beneficial.

2.2.3 Inelastic Electron Tunneling

In the electron tunneling described above, we considered the situation in which electrons are transferred between the electrodes without energy loss. This process is called elastic process. However, when oscillators, i.e. molecules, exist in the tunneling gap, tunneling electrons can interact with them. If the tunneling electron has enough energy to excite the vibration, it gives a part of energy to the vibrational excitation during the tunneling. In this process the energy of the electron is different between before and after the tunneling, which is called inelastic electron tunneling (IET) process. In the IET process a vibrational excitation opens an additional channel for tunneling electrons (Fig. 2.7a), giving rise to a substantial increase in the total current. The IET process was originally observed in metal-oxide-metal junctions where the molecules are included in the oxide layer [17, 18] as illustrated in Fig. 2.7b. We now turn to a bias voltage dependence of tunneling current in the tunnel junction including an oscillator (molecule) (Fig. 2.7c–e). Below a vibrational threshold ($eV < \hbar\omega$), the tunneling current increases almost linearly with ramping of the bias voltage, which is attributed to the elastic component. On the other hand, above the threshold ($eV = \hbar\omega$), electrons also tunnel via an inelastic channel. The probability passing the inelastic channel increases with the increase of bias voltage. Accordingly, the total current, including both elastic and inelastic components, shows a kink at $eV = \hbar\omega$ (Fig. 2.7c). The kink becomes the step and the peak in the conductance (dI/dV) (Fig. 2.7d) and the differential conductance (d^2I/dV^2) spectra (Fig. 2.7e), respectively. In general, since the fraction of the inelastic component is very small, like the conductance change is a few % (Fig. 2.7d). The peak in d^2I/dV^2 spectra indicates the vibrational energy and the intrinsic width, which is called IET spectra. The peak position, that is vibrational energy, corresponds to the signature of molecules and also represents the interaction with a substrate and surrounding conditions. Therefore we can chemically identify the species and speculate the structure of molecules in a tunnel junction. The peak width arises from three different contributions; the intrinsic width Γ of the vibration, the thermal broadening k_BT and the additional broadening depending on experimental conditions. The intrinsic width of vibration modes for adsorbates is expected to be the order of 1 meV or greater in surface species, with the peak profile described by a Lorentzian function. On the other hand, the thermal broadening is described by the Gaussian profile. Since the measurement of IETS is based on the lock-in detection, we also have to

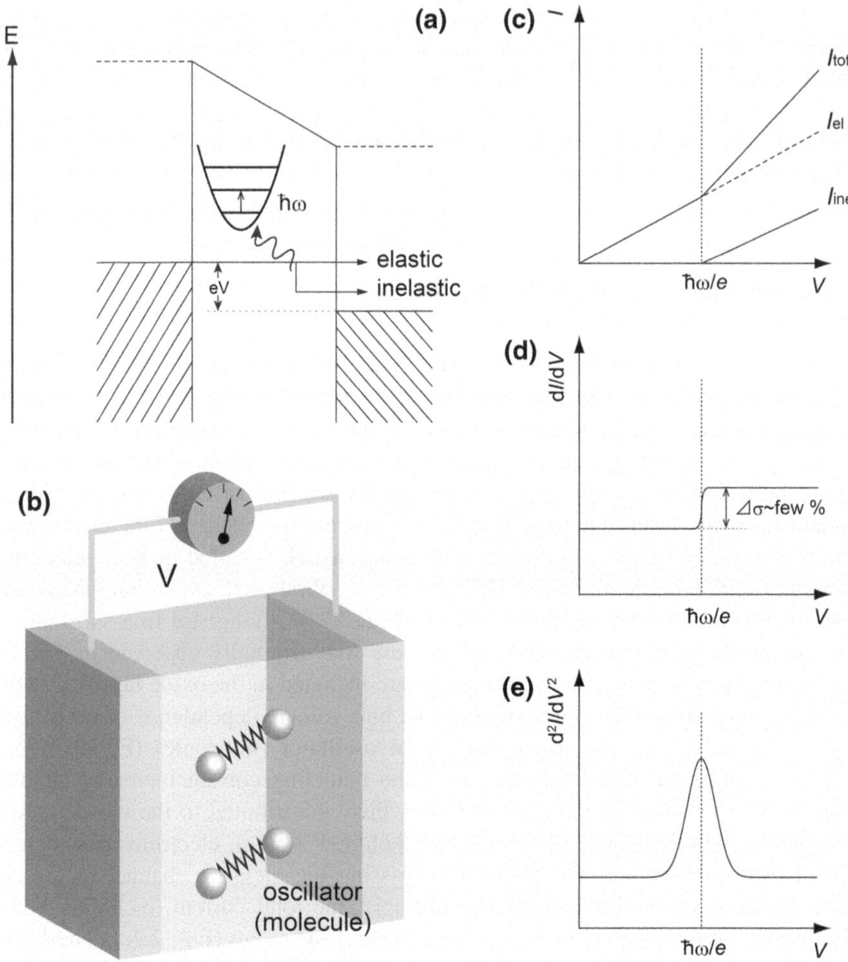

Fig. 2.7 **a** Schematic energy diagram of IET process. If an electron has enough energy to excite a vibration of molecules embedded in the tunnel junction, it gives a part of energy to the vibrational excitation and tunnels inelastically. This inelastic process opens an additional channel for the current flow. **b** Schematic illustration of a metal-oxide-metal junction including oscillators (molecules). **c** When the IET process occurs, a kink appears at the energy of molecular vibration $\hbar\omega/e$ in the I–V curve. **d–e** This kink becomes a step and a peak in the dI/dV and d^2I/dV^2 spectra, respectively, and the peak energy in d^2I/dV^2 spectrum corresponds to the vibration energy

take the broadening due to the modulation voltage into account. The resolution of IETS is usually governed by the thermal effect and modulation voltage and the total width in the d^2I/dV^2 signal is represented by the following Eqs. [19–22]

$$W = \sqrt{(1.7V_{\mathrm{rms}})^2 + (5.4k_{\mathrm{B}}T)^2 + \Gamma^2} \tag{2.13}$$

where V_{rms} is the modulation voltage, k_B is the Boltzmann constant, and T is the temperature. The typical resolution of IETS is expected to be 1-4 meV at low temperature ($\sim$5 K).

2.2.4 STM-IETS

The capability detecting molecular vibrations by combining STM and IETS was already mentioned in the 1980 s. Since tunneling electrons in STM pass through an area that is smaller than a single molecule, in principle, STM-IETS is able to measure molecular vibrations within a single molecule. The first result was reported in 1998 by Stipe et al. [23] and the C–H(D) stretch mode was clearly observed with the correct ratio of the isotope substitution. This result demonstrated the STM ability to be used as not only a microscope but also a chemically sensitive tool. At the same time, the result provided following questions. (i) Why only the C–H stretch mode is detected? (ii) What determines the shape of spectra? (i) is related to the selection rule, which is important and useful in vibrational spectroscopy for understanding the molecular states. (ii) is related to fundamental processes of a vibrational excitation/de-excitation of adsorbates. To answer these questions several experimental and theoretical attempts have been devoted in the last decade. The basic concept describing the elementary process of STM-IETS dates back to theoretical works reported in the late 1980s [24, 25]. Persson and Demuth first discussed the inelastic tunneling with the scheme of dipole scattering theory using Bardeen's formula for electric current [24]. After that, Persson and Baratoff showed that the resonant tunneling via adsorbate induced states is a dominant channel for inelastic process and the inelastic fraction of electronic current is associated with the vibrational damping rate due to the electron–hole pair excitation [25]. Figures 2.8a illustrates an STM junction consisted of a tip, substrate, and molecule chemisorbed on the substrate. The molecule forms the molecule-derived state ρ_a around the Fermi level E_F of the substrate, which is derived form a molecular orbital $|a>$ [16]. Such molecular-induced resonances frequently occur in the vicinity of E_F [26]. The width of the molecular state Γ is determined by the strength of the interaction between a molecule and an electronic state of an electrode. Tunneling electrons pass through the molecule-derived state ρ_a. When an electron has tunnel into ρ_a, it is temporary trapped in ρ_a for a time $t \sim \hbar/\Gamma$. If a tunneling electron has enough energy to excite a molecular vibration ($\hbar\omega$), it becomes possible to leave a vibrational excitation in the molecule and the electron inelastically-tunnels, which eventually opens an additional channel for the electron transport and results in the increase of a total current. This resonant tunneling model was experimentally examined by Kawai and her co-workers at RIKEN, Japan. They showed that the vibrational excitation depends on the spatial distribution of the molecular orbitals formed near the Fermi level of metal [27, 28].

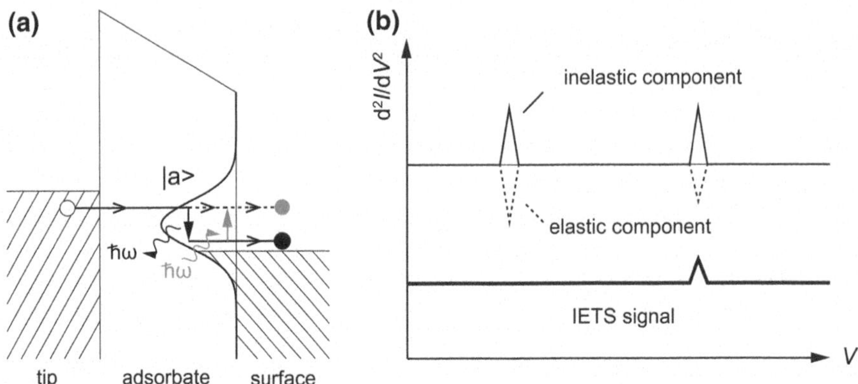

Fig. 2.8 **a** Schematic illustration of IET process via adsorbate-induced resonance state **a** The virtual emission (*black solid arrow*) and re-absorption (*gray solid arrow*) of a vibrational quantum also contribute to the direct elastic tunneling (*dashed arrow*). **b** Schematic IETS signals. The signal results from the sum of the elastic (*gray solid line*) and inelastic (*gray dashed line*) components

Although the resonant tunneling model provides an intuitive understanding of the process, it is not enough to answer the above questions. Several attempts to reproduce IETS spectra were performed by theorists [29–32], they provided the details of the underlying mechanism. Lorente et al. pointed out the importance of the contribution from not only the inelastic component but also the elastic component including the process that the electron emits and re-absorbs the virtual vibration as illustrated in Fig. 2.8 [29]. They carried out the density functional theory calculations of the vibrational IET process of acetylene on Cu(100) and calculated elastic and inelastic components based on a many-body generalization of the Tersoff-Hamann theory. The results explained why only the C–H stretch mode could be observed in the STM-IETS. Ueba and his co-workers presented a description of STM-IETS signal using the Keldysh Green's function method for the molecule-induced resonance model. The elastic and inelastic tunneling currents are derived on the basis of the Anderson Hamiltonian coupled to a vibrational degree of freedom where effects of vibrational damping rate on the inelastic current were taken into consideration [31, 32].

The important finding of the theoretical efforts is that the modification of the elastic component always gives a reduction of tunneling current, while the inelastic contribution gives an increase. In summary, the total change of tunneling current is determined by the competition between the variations of elastic and inelastic components, and a vibration mode in which the elastic or inelastic component sufficiently overcome another contribution can be observed (Fig. 2.8b). These approaches explained why only limited vibrational modes are detected in the STM-IETS, e.g., only C–H (C-D) stretch mode for C_2H_2. The competition between the negative elastic and positive inelastic components can also lead to even negative conductance change. Such a decrease in conductance is predicted when a molecular state is resonant with the Fermi energy [33]. In fact, the decrease of the total current was observed in a single oxygen molecule adsorbed on Ag(110) [34, 35].

The spectral shape of IETS is of importance in understanding elementary processes of a vibrational excitation/de-excitation of adsorbates [36]. However it is not straightforward to estimate the vibration broadening, which includes fundamental properties of vibration, like lifetime and dissipation processes, in STM-IETS. Lauhon and Ho systematically studied the effect of temperature and the bias modulation on the vibrational peak width, intensity and line shape for the C–H stretch mode of a single acetylene molecule on Cu(001) [37]. They concluded that the IETS spectra are less influenced by vibrational damping because of the use of large bias modulation than the vibrational broadening. This may suppress the intrinsic features in the IETS spectra.

The selection rule of vibrational spectroscopy is important and useful to understand the molecular structure and orientation on surfaces. Several attempts to elucidate the selection rule were carried out from both of theoretical [33, 38] and experimental [39] aspects. Although a rigorous selection rule of the STM-IETS is still open question, Troisi and Ratner [40], and Paulsson et al. [33] proposed *propensity rules* that give useful insights to the geometric and electronic structure of nano-scale junctions containing atomic wires or molecules. Troisi and Ratner calculated the IETS signal of the junction in which two electric leads are bridged by molecules. After that, Paulsson extended the model to the unified description for electron transport systems within nano-junctions including an STM configuration. Figure 2.9 shows a phase diagram of the conductance change at a vibrational excitation. Here it is assumed that a single electronic level (ε_0) is coupled with two electric leads and a localized vibration (one-level model). The phase diagram is characterized by the symmetric factor α (vertical axis) and the transmission probability τ of electron in Landauer model (transverse axis). The crossover from a decrease to an increase (boundary depicted dashed line) stems from crossing the resonance. For this model the maximal transition corresponds to the on-resonance case, which given by $\tau_{max} = 4\alpha/(1 + \alpha)$. The symmetric factor is the ratio of the coupling between the two electric leads ($\Gamma_{R/L}$). The asymmetric configuration of STM junction gives $\alpha \approx 0$ while the symmetric junction bridged by a molecule (molecular wire) $\alpha \approx 1$. On the other hand, very low τ is expected in the STM and molecular wires, while atomic wire exhibit $\tau \approx 1$. The propensity rule can be examined using the extensive DFT calculations as described in Ref. [33].

For STM-IETS measurement, the high stability of STM and low extrinsic noise level of the experimental setup are required, which was the main obstacle to realize STM-IETS. Since the measurement relies on the lock-in technique, it takes a few minutes to obtain one d^2I/dV^2 spectrum and the feedback loop keeping z position of the STM tip is needed to be opened during the bias sweep. The relative distance between the STM tip and adsorbate must be maintained to be constant to obtain a correct signal. As described in Sect. 2.2.2, a displacement between the tip and adsorbate causes a significant variation in tunneling current even if it is very small, which suppresses the vibrational signal from the adsorbate. To minimize this displacement (mainly caused by thermal drift) and thermal broadening of a molecular

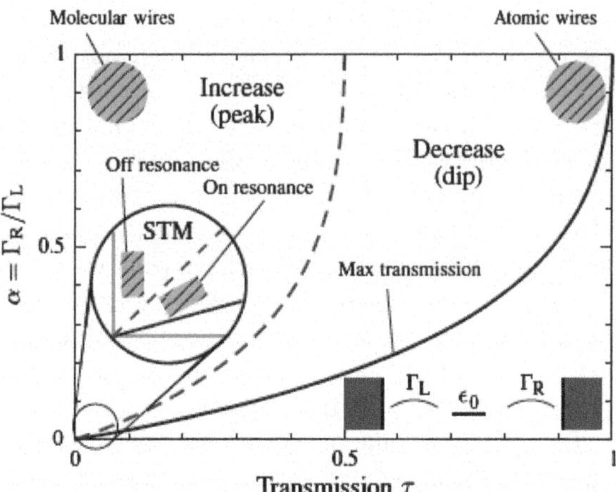

Fig. 2.9 Phase diagram for a one-level model (*inset*) illustrating the sign of the conductance change at the onset of phonon emission. At a given asymmetry factor the elastic transmission has an upper boundmax (*black line*), and the inelastic conductance change undergoes a sign change at crossover max = 2 (*dashed line*). Reprinted with permission from Ref. [33]. Copyright 2008, American Physical Society

vibration, a cryogenic condition is usually requested. Additionally, adsorbates have to be stationary under the STM tip during the measurement. When a molecule motion is induced by thermal activation or the excitation of a particular vibrational mode with tunneling electron, it becomes difficult to measure STM-IETS. However the latter phenomena can be used as an alternative vibrational spectroscopy of STM, which is discussed in the next section.

2.2.5 STM-Induced Motion and Reaction of Adsorbates

Tunneling electrons from STM can be used to induce motions and reactions of individual adsorbates. Several kinds of molecular motions and reactions have been characterized at the single molecule level in the past decade. The trailblazing works were reported by Stipe, Rezaei, and Ho in the late 1990s [41–43]. They induced and imaged of a rotation and dissociation of single oxygen molecules adsorbed on a Pt(111) surface. The mechanism was rationalized by intramolecular vibrational excitations via resonant inelastic electron tunneling. Since their pioneering efforts, various kinds of single-molecule reactions, e.g., desorption of ammonia molecule [44], bond-selective reaction of ammonia molecule [45–47], deprotonation of acetylene, benzene [48, 49] and *trans*-2-butene molecule [50, 51], conformational switching of porphyrin-based molecule [52] have been reported. After these findings, Kawai and her co-workers at RIKEN introduced a new concept of vibrational spectroscopy with STM, the so-called inelastic electron tunneling action spectroscopy (IET-AS) or STM action spectroscopy (STM-AS) [53]. They measured the reaction yield of a vibrationally induced the configurational change of *cis*-2-butene on Pd(110) as a function of applied bias voltage [Fig. 2.10]. It was found that rapid increases of reaction yield were directly related to the energy of vibrational modes of *cis*-2-butene. In most cases the reaction yield

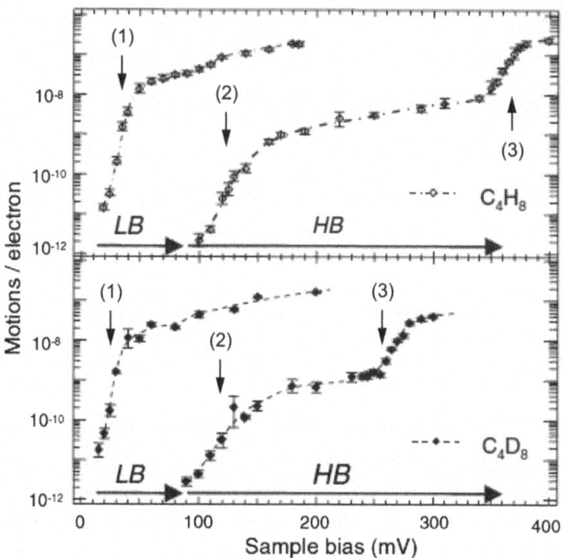

Fig. 2.10 STM-AS for *cis*-2-Butene on Pd(110). Abrupt increases of the motion rate are associated with molecular vibrations. [The number of arrow, mode, energy measured by EELS] = [(1), Pd–C stretch, 21, 36 meV], [(2), C–C stretch, 125 meV], [(3), C–H stretch, 357 meV]. Reprinted with permission from Ref. [53]. Copyright 2005, American Physical Society

(rate) is increased by a factor of 10-100 after a specific mode excitation. Interestingly, the STM-AS of *cis*-2-butene showed the excitation of ν(M-C), $\delta(CH_3)$, ν(C–C), and ν(C–H) modes but only ν(M-C) and ν(C–H) are detected in the STM-IETS. Therefore the STM-AS can be considered as a complementary method of the STM-IETS to conduct the chemical identification with the STM. The STM-AS is applied to identify several chemical adsorbates [54–63]. More recently, the practical method to extract vibrational information from the STM-AS was developed by Motobayashi et al. [64] at RIKEN, which can be generally applied to vibrationally mediated motions and/or reactions of adsorbates. The experimental techniques to obtain the rates of reaction and motion of a single molecule are mentioned in next chapter.

I now turn to the mechanism of STM-induced motions and reactions. The process resulting in the molecular motion or reaction can be classified into the direct and indirect process. The direct process is quite simple where tunneling electron from STM excites a vibration mode that is directly associated with the coordinate of a motion or reaction. In this case we need to consider a simple vibration ladder climbing model as shown in Fig. 2.11a. The questions are the barrier height E_B and the number of vibrational levels in the potential well, which can be determined by investigating the number of electrons, n, required inducing the motion or reaction. The rate is given by the so-called power law, $R \propto I^n$, here I is the tunneling current. If a tunneling electron has enough energy to overcome the barrier, the process proceeds with one-electron ($n = 1$; solid arrow). On the other hand, if the energy is not sufficient, multiple electrons are required to overcome the barrier (dashed arrows). The direct process was observed in Xe atom switching between a tip and a Ni(110) surface [65] and the dissociation of single oxygen molecules on a Pt(111) surface [41]. The power law dependence of $R \propto$

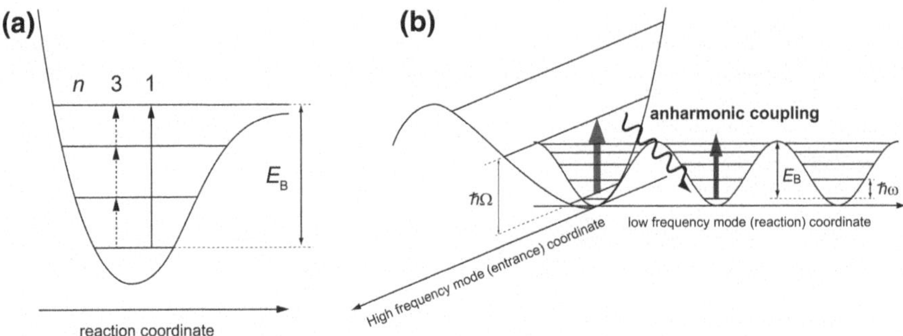

Fig. 2.11 Schematic illustration of the energy diagram of **a** direct and **b** indirect process

$I^{4.9\pm0.2}$ was found for the Xe switching, suggesting that the transfer of the Xe atom takes place through over-barrier process and the stepwise climbing of a vibrational ladder of the Xe-surface bond excitation is involved. The theoretical description of the direct process was reported by Walkup et al. [66] and Gao et al. [67]. The switching was described as the energy barrier climbing process between two potential wells modeled by truncated harmonic oscillators, where the higher vibrational states are populated by the excitation of tunneling electrons. On the other hand, for O_2 molecule dissociation on a Pt(111) surface, it was revealed a power low dependence of $N \sim 1$, 2, and 3 for the applied voltage of 0.4, 0.3, and 0.2 V, respectively. The excitation of the O–O stretching ($\hbar\Omega = 87$ meV) is associated with the dissociation and the potential barrier of 0.35–0.38 eV with 5 vibrational levels in the well was calculated using a truncated harmonic oscillator. Thus, the coherent multiple jump of vibrational ladders in a one-electron scattering dominates the dissociation at 0.4 V to overcome the barrier, while the step-by-step ladder climbing through two (three)-electron process makes it possible to break the O–O bond at 0.3 (0.2) V.

On the other hand, if the motion or reaction of adsorbates is triggered by the mode that is not directly coupled with its coordinate (indirect process), the intramolecular energy transfer have to be taken into account. This indirect process was discovered in the rotation of oxygen and acetylene molecule [42, 43]. For instance, the threshold energy to induce the rotation of acetylene molecule on Cu(100) coincides with the C–H stretch excitation [43]. Obviously the C–H stretch mode is not the reaction coordinate of the rotation of acetylene. The mechanism was rationalized by the intramolecular energy transfer via anharmonic coupling between the high and low frequency modes. We now consider the fate of a molecular vibration excited via IET process. After the excitation, the excited state rapidly damps to the ground state via intramolecular energy transfer, surface phonon excitation, or electron–hole-pair (EHP) excitations on a time scale of femtosecond to picosecond. Among them the intramolecular energy transfer can eventually causes the molecular motions and reactions. Figure 2.11b illustrates the concept of the intramolecular energy transfer. When a high frequency (HF) mode is excited via IET process, its energy relaxes to a lower frequency modes associated

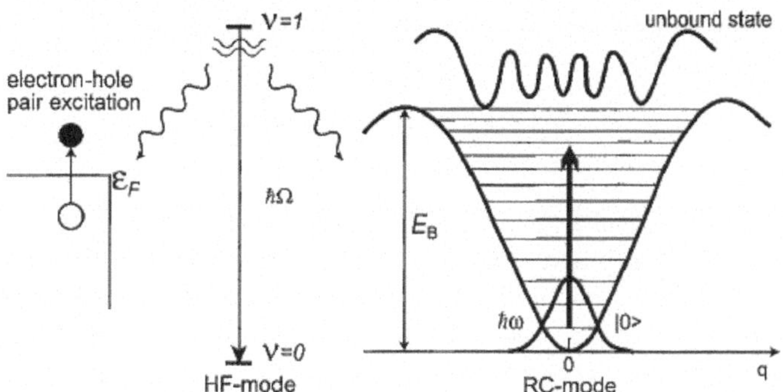

Fig. 2.12 Schematic illustration of single-electron process. The decay of the high-frequency (HF) mode with the energy of $\hbar\Omega \gg E_B$ excites the low frequency reaction coordinate mode from the ground state localized at the bottom of the RC-potential well to the unbound state just above the top of the barrier, in competing with EHP excitations. Reprinted with permission from Ref. [69]. Copyright 2005, American Institute of Physics

with a specific reaction coordinate (RC mode) via anharmonic coupling. If a sufficient energy to surmount the barrier is accumulated in the reaction coordinate, the motion or reaction takes place. This mechanism was clearly demonstrated in the lateral hopping of carbon monoxide (CO) molecules adsorbed on a Pd(110) surface by Komeda et al. [68]. They showed a quantitative description of the anharmonic coupling between the HF and RC modes. Their results highlighted microscopic processes of the intramolecular energy transfer within single molecules adsorbed on metal surfaces. It was found the motion or reaction rate of indirect process also follows the power low [42]. Ueba et al. developed a theory quantifying elementary processes of adsorbate motions induced via intermolecular mode coupling [69], where they discussed one- and two-electron processes for IET-induced motions based on rate equations describing the evolution of a vibrational population of an HF mode and of an RC mode. They discussed the reaction rate in terms of inelastic tunneling rates, vibrational relaxation rates, and intermode anharmonic coupling rates. Figure 2.12 shows the proposed model for one-electron process, where anharmonic mode coupling between the HF mode (excited by tunneling electron) and the RC mode results in the excitation of the RC mode to a level above E_B. For one-electron process the vibrational energy of HF mode $\hbar\Omega$ is larger than E_B. In the model of Fig. 2.12 the reaction rate R_{RC} can be obtained as

$$R_{RC} = \frac{\tau_v}{\tau_{v,RC}} \eta_v I \qquad (2.14)$$

where $\tau_v/\tau_{v,RC}$ is the ratio of the excitation rate of the RC mode above the potential barrier $E_B(\ll \hbar\Omega)$, and the decay rate of the v mode into low-energy excitations in the substrate (EHP excitations and bulk phonons), η_v and I are the inelastic-tunneling fraction and the tunneling current, respectively. The calculation of $\tau_{v,RC}$ depends on the model of anharmonic coupling between the HF and RC modes. In the case of a CO hopping on Pd(110) the HF mode (C–O stretch) is transferred to transnational or rotational energy along the reaction coordinate, in competition with the fast vibrational relaxation through EHP excitation, gives

$$\frac{1}{\tau_{v,\mathrm{RC}}} \approx \frac{1}{\tau_v} \left(\frac{\hbar\, \delta\omega}{E_\mathrm{B}} \right) \chi^{3/2} e^{-2\alpha} \tag{2.15}$$

$\delta\omega$ represents the anharmonic coupling between the HF and RC modes, and $\chi = E_\mathrm{B}/\hbar\omega$. This theory agrees in the order of magnitude with the experimental result of a CO hopping rate on Pd(110). It is noted that this theory could explain why CO hopping is not observed on Cu(110) on contrast to Pd(110), which follows from the small $\delta\omega$ and large χ [68–70].

In short STM-AS provides elementary processes of adsorbate motions and reactions as well as fruitful information about vibrational excitations. Two different mechanisms have been characterized: i) direct process in which the reaction coordinate mode is directly excited by tunneling electrons, ii) indirect process that involves energy transfer between vibrational modes via anharmonic coupling.

I focused on the vibrational mediated reactions of adsorbates. However, the reactions can be also induced by electronic excitation or electron (hole) attachment with STM, which takes place at the bias voltage higher than $\sim$1 eV [71–74]. The electronic excitation occurs when tunneling electrons couple with electronic states of the atoms or molecule.

STM can induce and image a variety of fascinating phenomena at the single molecule level. Proper understanding of current-driven events in STM junctions is particular interest in molecular-scale electronics because it is directly related to the conductivity, heating and current-induced failure. From a theoretical perspective, current-driven dynamics in STM junctions includes the challenging subjects to describe non-equilibrium and non-adiabatic phenomena under the bias voltage and the dissipative effects of the electrodes.

2.2.6 Single Atom and Molecule Manipulation

An important application of tip-adsorbate interaction is the manipulation of single atoms and molecules on surfaces, which offers a precise positioning of individual adsorbates and assembly of nano-scale structures at the spatial limit. The interaction between the adsorbate and the surface is also important in the STM manipulation as well as the force working between the tip and the adsorbate. While a strong adsorbate–surface interaction makes the manipulation impossible, a weak interaction results in an uncontrollable situation.

The first example of the atom manipulation was reported by Eigler and Schweizer in 1990 [75]. They transferred Xe atoms back and forth reversibly between the tip and the surface and positioned them in fully controlled fashion on a Ni(110) surface. Figure 2.13a shows a schematic illustration of this manipulation that is called "vertical manipulation". It is also found that single atoms can be

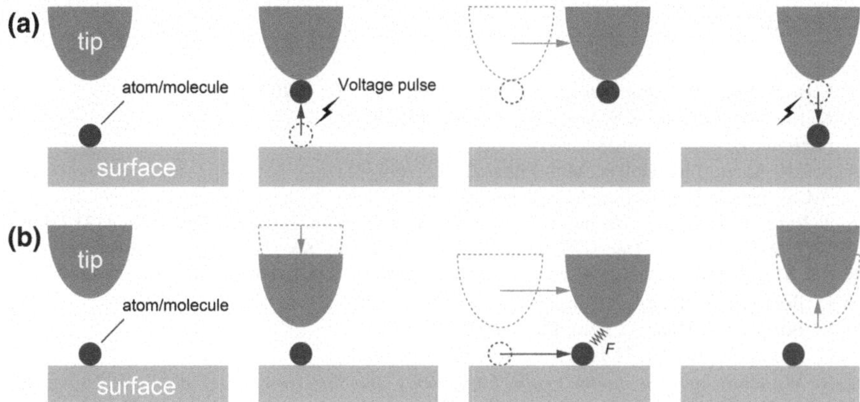

Fig. 2.13 Schematic illustration of vertical **a** and lateral manipulation

moved along predetermined paths of an STM tip [76]. This process is illustrated in Fig. 2.13b, which is called "lateral manipulation". Since the proximity of an STM tip causes the strong field, the tip gives a finite force on an adsorbate [77]. This force involves both van der Waals and electrostatic contributions, which can be adjusted by controlling the tip-surface separation and the bias voltage. Bartels, Meyer and Rieder clearly demonstrated several kinds of lateral manipulation modes depending on tunneling parameters [78]. The lateral manipulation also applied to large molecules including molecular machines [79–82].

Understanding the force acting between the tip and the adsorbate is fundamental importance in STM manipulations. However it is quite hard to measure the force in STM. Recently, Ternes et al. measured the vertical and lateral forces exerted on individual adsorbed atoms or molecules by the tip [83]. They detected such forces using the simultaneous operation of STM and atomic force microscope (AFM). It was found that the force moving cobalt (Co) on Pt(111) and Co on Cu(111) required a lateral force of 210 and 17 pN, respectively. The emergence of STM/AFM paves a way to understanding the driving mechanism.

Although the STM manipulation has been well established so far, AFM can also be used for the manipulation single atoms and molecules. Recent progress of the non-contact AFM enables to manipulate atoms at semiconducting surfaces [84], even at room temperature [85, 86].

Even though the construction of nano-scale structure using STM manipulation is inherently quite slow process, it gives fascinating opportunities to investigate physical and chemical processes in desirable model systems at the single atom/molecule limit.

References

1. J. Brickmann, H. Zimmermann, J. Chem. Phys. **50**, 1608 (1969)
2. K. Liu, J.D. Gruzan, R.S. Saykally, Science **271**, 929 (1996)
3. F. Hund, Z. Phys. **43**, 803 (1927)
4. A.C.P. Alves, J.M. Hollas, Mol. Phys. **23**, 927 (1972)
5. A.C.P. Alves, J.M. Hollas, Mol. Phys. **25**, 1305 (1973)
6. T. Baba, T. Tanaka, I. Morino, K.M.T. Yamada, K. Tanaka, J. Chem. Phys. **110**, 4131 (1999)
7. W.F. Rowe Jr, R.W. Duerst, E.B. Wilson, J. Am. Chem. Soc. **98**, 4021 (1976)
8. S.L. Baughcum, Z. Smith, E.B. Wilson, R.W. Duerst, J. Am. Chem. Soc. **106**, 2260 (1984)
9. D. Borgis, J.T. Hynes, J. Chem. Phys. **94**, 3619 (1991)
10. D. Borgis, J.T. Hynes, Chem. Phys. **170**, 315 (1993)
11. G. Binnig, H. Rohler, Helv. Phys. Acta **55**, 725 (1982)
12. R. Wiesemdanger, *Scanning Probe Microscopy and Spectroscopy* (Cambridge University Press, Cambridge, 1994)
13. J.A. Strocsio, W.J. Kaiser, T. Lucatorto, M.D. Graef, *Scanning Tunneling Microscopy* (Academic Press, Methods of Experimental Physics, 1994)
14. C. Bai, *Scanning tunneling microscopy and its applications* (Springer, Berlin, 2000)
15. C.J. Chen, *Introduction to Scanning Tunneling Microscopy* (Oxford University Press, Monograph on the Physics and Chemistry of Materials, 2007)
16. J. Tersoff, D.R. Hamann, Phys. Rev. Lett. **50**, 1998 (1983)
17. R.C. Jaklevic, J. Lambe, Phys. Rev. Lett. **17**, 1139 (1966)
18. J. Lambe, R.C. Jaklevic, Phys. Rev. **165**, 821 (1968)
19. P.K. Hansma, Phys. Rep. **30C**, 145 (1977)
20. W.H. Weinberg, Annu. Rev. Phys. Chem. **29**, 115 (1978)
21. T. Wolfram (ed.), *Inelastic Electron Tunneling Spectroscopy* (Springer-Verlag, Berlin, 1978)
22. P.K. Hansma (ed.), *Tunneling Spectroscopy* (Plenum, NewYork, 1982)
23. B.C. Stipe, M.A. Rezaei, W. Ho, Science **280**, 1732 (1998)
24. B.N.J. Persson, J.E. Demuth, Solid State Commun. **57**, 769 (1986)
25. B.N.J. Persson, A. Baratoff, Phys. Rev. Lett. **59**, 339 (1987)
26. N.D. Lang, A.R. Williams, PRB **18**, 616 (1977)
27. M. Kawai, T. Komeda, Y. Kim, Y. Sainoo, S. Katano, Phil. Trans. R. Soc. Lond. A **362**, 1163 (2004)
28. M. Ohara, Y. Kim, S. Yanagisawa, Y. Morikawa, M. Kawai, Phys. Rev. Lett. **100**, 136104 (2008)
29. N. Lorente, M. Persson, Phys. Rev. Lett. **85**, 2997 (2000)
30. N. Mingo, K. Makoshi, Phys. Rev. Lett. **84**, 3694 (2000)
31. T. Mii, S. Tikhodeev, H. Ueba, Surf. Sci. **502**, 26 (2002)
32. T. Mii, S.G. Tikhodeev, H. Ueba, Phys. Rev. B **68**, 205406 (2003)
33. M. Paulsson, T. Frederiksen, H. Ueba, N. Lorente, M. Brandbyge, Phys. Rev. Lett. **100**, 226604 (2008)
34. J.R. Hahn, H.J. Lee, W. Ho, Phys. Rev. Lett. **85**, 1914 (2000)
35. M. Alducin, D. Sánchez-Portal, A. Arnau, N. Lorente, Phys. Rev. Lett. **104**, 136101 (2010)
36. M. Galperin, M.A. Ratner, A. Nitzan, J. Chem. Phys. **121**, 11965 (2004)
37. L.J. Lauhou, W. Ho, Rev. Sci. Instrum. **72**, 216 (2001)
38. N. Lorente, M. Persson, L.J. Lauhon, W. Ho, Phys. Rev. Lett. **86**, 2593 (2001)
39. N. Okabayashi, M. Paulsson, H. Ueba, Y. Konda, T. Komeda, Phys. Rev. Lett. **104**, 077801 (2010)
40. A. Troisi, M.A. Ratner, J. Chem. Phys. **125**, 214709 (2006)
41. B.C. Stipe, M.A. Rezaei, W. Ho, S. Gao, M. Persson, B.I. Lundqvist, Phys. Rev. Lett. **78**, 4410 (1997)
42. B.C. Stipe, M.A. Rezaei, W. Ho, Science **279**, 1907 (1998)
43. B.C. Stipe, M.A. Rezaei, W. Ho, Phys. Rev. Lett. **81**, 1263 (1998)

44. L. Bartels, M. Wolf, T. Klamroth, P. Saalfrank, A. Kühnle, G. Meyer, K.-H. Rieder, Chem. Phys. Lett. **313**, 544 (1999)
45. N. Lorente, J.I. Pascual, Phil. Trans. **362**, 1227 (2004)
46. J.I. Pascual, N. Lorente, Z. Song, H. Conrad, H.-P. Rust, Nature **423**, 525 (2003)
47. N. Lorente, J.I. Pascual, H. Ueba, Surf. Sci. **593**, 122 (2005)
48. L.J. Lauhon, W. Ho, Surf. Sci. **451**, 219 (2000)
49. L.J. Lauhon, W. Ho, J. Phys. Chem. A **104**, 2463 (2000)
50. Y. Kim, T. Komeda, M. Kawai, Surf. Sci. **502**, 7 (2002)
51. Y. Kim, T. Komeda, M. Kawai, Phys. Rev. Lett. **89**, 126104 (2002)
52. F. Moresco, G. Meyer, K.-H. Rieder, H. Tang, A. Gourdon, C. Joachim, Phys. Rev. Lett. **86**, 672 (2001)
53. Y. Sainoo, Y. Kim, T. Okawa, T. Komeda, H. Shigekawa, M. Kawai, Phys. Rev. Lett. **95**, 246102 (2005)
54. J.R. Hahn, W. Ho, J. Chem. Phys. **131**, 044706 (2009)
55. M. Parschau, D. Passerone, K.-H. Rieder, a. H. J. Hug, and K.-H. Ernst. Angew. Chem. Int. Ed. **48**, 4065 (2009)
56. S. Katano, Y. Kim, Y. Kagata, M. Kawai, J. Phys. Chem. C **113**, 19277 (2009)
57. S. Katano, Y. Kim, Y. Kagata, M. Kawai, J. Phys. Chem. C **114**, 3003 (2010)
58. H.-J. Shin, J. Jung, K. Motobayashi, S. Yanagisawa, Y. Morikawa, Y. Kim, M. Kawai, Nature Mater. **9**, 442 (2010)
59. V. Simic-Milosevic, M. Mehlhorn, K.-H. Rieder, J. Meyer, K. Morgenstern, Phys. Rev. Lett. **98**, 116102 (2007)
60. V. Simic-Milosevic, K. Morgenstern, J. Am. Chem. Soc. **131**, 416 (2009)
61. V. Simic-Milosevic, J. Meyer, K. Morgenstern, Angew. Chem. Int. Ed. **48**, 4061 (2009)
62. V. Simic-Milosevic, J. Meyer, K. Morgenstern, Phys. Chem. Chem. Phys. **10**, 1916 (2008)
63. W. Wang, X. Shi, M. Jin, C. Minot, M.A. Van Hove, J.-P. Collin, N. Lin, ACS Nano **4**, 4929 (2010)
64. K. Motobayashi, Y. Kim, H. Ueba, and Maki Kawai. Phys. Rev. Lett. **105**, 076101 (2010)
65. D.M. Eigler, C.P. Lutz, W.E. Rudge, Nature **352**, 600 (1991)
66. R.E. Walkup, D.M. Newns, P. Avouris, Phys. Rev. B **48**, 1858 (1993)
67. S. Gao, M. Persson, B.I. Lundqvist, Phys. Rev. B **55**, 4825 (1997)
68. T. Komeda, Y. Kim, Maki Kawai, B. N. J. Persson, and H. Ueba, Science, 295, 2055 (2002)
69. H. Ueba, T. Mii, N. Lorente, B.N.J. Persson, J. Chem. Phys. **123**, 084707 (2005)
70. B.N.J. Persson, H. Ueba, Surf. Sci. **502**, 18 (2002)
71. A.J. Mayne, G. Dujardin, G. Comtet, D. Riedel, Chem. Rev. **106**, 4355 (2006)
72. A. van Houselt, H.J.W. Zandvliet, Rev Mod Phys **82**, 1593 (2010)
73. P. Liljeroth, J. Repp, G. Meyer, Science **317**, 1206 (2007)
74. T.-C. Shen, C. Wang, G.C. Abeln, J.R. Tucker, J.W. Lyding, Ph Avouris, R.E. Walkup, Science **268**, 1590 (1995)
75. D.M. Eigler, E.K. Schweizer, Nature **344**, 524 (1990)
76. J.A. Stroscio, D.M. Eigler, Science **254**, 1319 (1991)
77. U. Dürig, O. Züger, D.W. Pohl, J. Microscopy **152**, 259 (1988)
78. L. Bartels, G. Meyer, K.-H. Rieder, Phys. Rev. Lett. **79**, 697 (1997)
79. P.H. Beton, A.W. Dunn, P. Moriarty, Appl. Phys. Lett. **67**, 1075 (1995)
80. J.K. Gimzewski, C. Joachim, R.R. Schlittler, V. Langlais, H. Tang, I. Johannsen, Science **281**, 531 (1998)
81. F. Moresco, G. Meyer, K.-H. Rieder, H. Tang, A. Gourdon, C. Joachim, Phys. Rev. Lett. **87**, 088302 (2001)
82. L. Grill, K.-H. Rieder, F. Moresco, G. Rapenne, S. Stojkovic, X. Bouju, C. Joachim, Nat. Nanotech. **2**, 95 (2007)
83. M. Ternes, C.P. Lutz, C.F. Hirjibehedin, F.J. Giessibl, A.J. Heinrich, Science **319**, 1066 (2008)
84. N. Oyabu, Y. Sugimoto, M. Abe, Ó. Custance, S. Morita, Nanotechnology **16**, S112 (2005)
85. Y. Sugimoto et al., Nat. Mater. **4**, 156 (2005)
86. Y. Sugimoto et al., Phys. Rev. Lett. **98**, 106104 (2007)

Chapter 3
Experiments

Abstract I describe experimental details in this chapter. All experiments were conducted under ultra-high vacuum (UHV) conditions. In general, since single-crystalline surfaces of semiconductor and metal are very reactive, they are readily contaminated by H_2O, O_2, etc. in atmospheric conditions. Therefore, UHV conditions ($<10^{-10}$ Torr) are essential to avoid contaminations from such gases and keep well-defined conditions at the atomic level. Principle and technical details of UHV can be found in some practical books [1–3]. A single-crystalline Cu(110) surface was used as an atomically well-defined template throughout this thesis. As described in the previous chapter, the structure of water layer on metal surfaces is determined by the competition between water–water and water-substrate interactions. In this context, a Cu(110) surface is quite unique because several unusual structures, including a one-dimensional chain [4, 5], an intact two-dimensional network [6], and some partially dissociated overlayers [7–9] are formed. This feature is presumably attributed to the quasi one-dimensional atomic arrangement of Cu(110). The experimental set up, sample preparation, and STM measurement are summarized in Sects. 3.1, 3.2, and 3.3, respectively. Additionally, I focus on the time-resolved measurement using STM in Sect. 3.4.

Keywords Ultra-high vacuum conditions · Low temperature · Single crystalline surfaces · A Cu(110) surface · Scanning tunneling microscope

3.1 Experimental Apparatus

The basic set-up of the STM is shown in Fig. 3.1. There are two separated chambers for preparing a single-crystalline surface and for STM measurement at low-temperature. The tall part in the STM chamber is a refrigeration tank (cryostat). Temperature of 78 and 6 K can be achieved by filling the cryostat with liquid N_2

T. Kumagai, *Visualization of Hydrogen-Bond Dynamics*, Springer Theses,
DOI: 10.1007/978-4-431-54156-1_3,

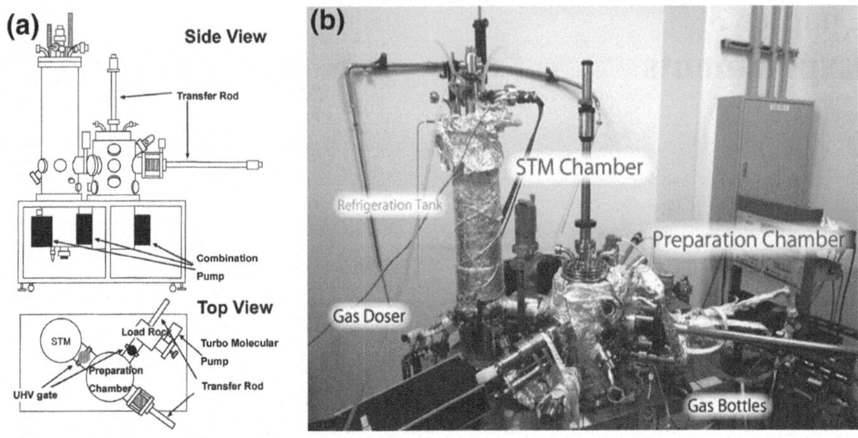

Fig. 3.1 **a** Schematic illustration of the LT-STM system. **b** Overview picture of the LT-STM

and He, respectively. The STM head is cooled by the contact to the Cu block on the bottom of the inner cryostat. During the measurement, the STM head is suspended by springs for the vibrational isolation. Cu bundles were used to connect parts of the head for a rapid thermal equilibrium. The vacuum system consists of sputter ion pumps, Ti sublimation pumps, a turbo molecular pump and a rotary pump. UHV conditions can be obtained by baking the chamber for 3 or 4 days at $\sim$400 K, and are kept using ion pumps and titanium sublimation pumps. The cryostat in the STM chamber also works as a cryo-pump. STM measurements were conducted below 5×10^{-11} Torr. In this vacuum the contamination of surfaces can be avoided during the experimental time scale.

3.2 Sample Preparation

A Cu(110) surface was cleaned by repeated cycles of Ar^+ sputtering and annealing up to $\sim$800 K under UHV conditions to get an atomically clean surface. Ion sputtering is a conventional and effective cleaning method of metal surfaces [1–3]. Since the ion bombardment causes the degradation of the surface structure, subsequent annealing is essential to restore the crystalline surface and eliminate embedded and adsorbed Ar atoms. Figure 3.2 shows a high-resolution STM image of the clean Cu(110) surface that has an anisotropic arrangement as illustrated in the STM image. Cu atomic rows run along the $[1\bar{1}0]$ direction. The quasi 1-D geometory makes the dynamics of molecules anisotropic, giving rise to unique directional growth of water layers [4–9].

In order to obtain isolated single water molecules, the surface was exposed to high-purely H_2O or D_2O gases below 20 K via a tube doser positioned $\sim$1 cm apart from the surface. The exposure amount was controlled using a leak valve.

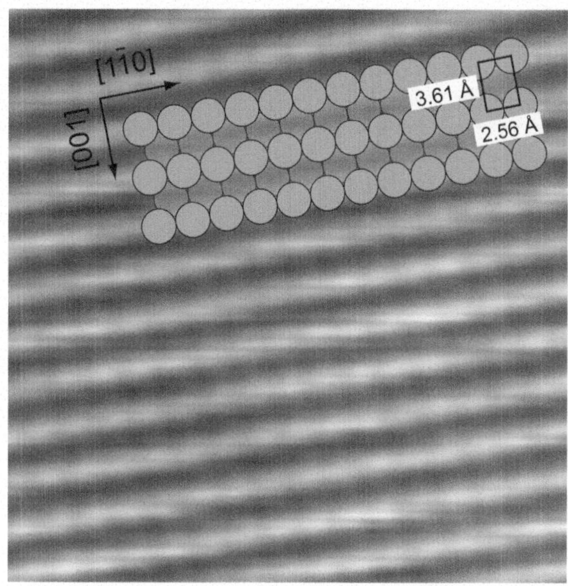

Fig. 3.2 STM image of the Cu(110) clean surface. The atomic rows of Cu running along [1$\bar{1}$0] direction are observed

I conducted most of the experiments at extremely low coverages, where water molecules exist mainly as an isolated monomer or dimer.

3.3 STM Measurement

All STM measurements were carried out at 6 K. A bias voltage is applied to a sample, which is denoted as V_s. Tunneling current denoted as I_t, is detected after conversion to voltage by a current amplifier.

I used a tungsten (W) tip as a probe for all STM measurements. W is commonly used material for STM tip under UHV conditions because W has a mechanical stiffness and a relatively flat electron density of states (DOS) near the E_F that makes it easy to interpret an obtained STM image. W tips were fabricated by electrochemical etching of W wires in NaOH solution. STM tips fabricated in atmosphere are usually covered by oxide layers. To remove the oxide layers and impurities, I performed in situ tip cleaning by applying a high bias voltage which induces a field evaporation of atoms on the tip apex [10] and restructuring of the tip itself due to a non-uniform electric field [11].

All STM images presented in this thesis are acquired in constant current mode. The STM tip is moved from the top to bottom (Raster scan) while keeping the tip height (z) to give a constant tunneling current (at set current, I_t) via the feedback loop as illustrated in Fig. 3.3. The variation in surface LDOS causes a change in tunneling current and then z is adjusted with a given feedback frequency. Consequently the record of dz represents surface corrugations. Due to its straightforward

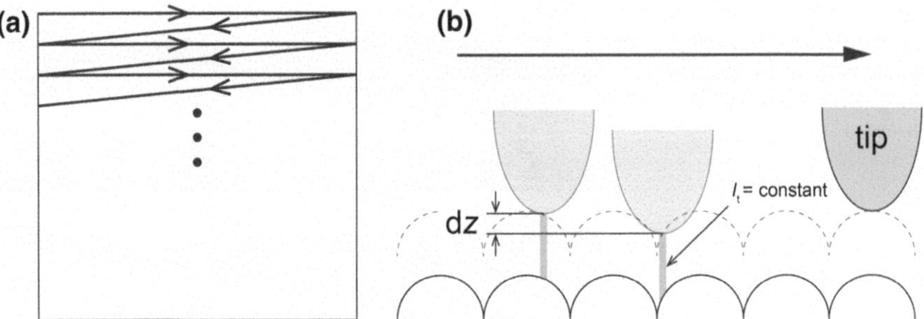

Fig. 3.3 Schematic illustration of STM measurement. **a** The STM tip is scanned from top to bottom with raster scan. **b** In constant current mode the relative distance between the STM tip and surface is controlled via feedback loop to keep the tunneling current constant

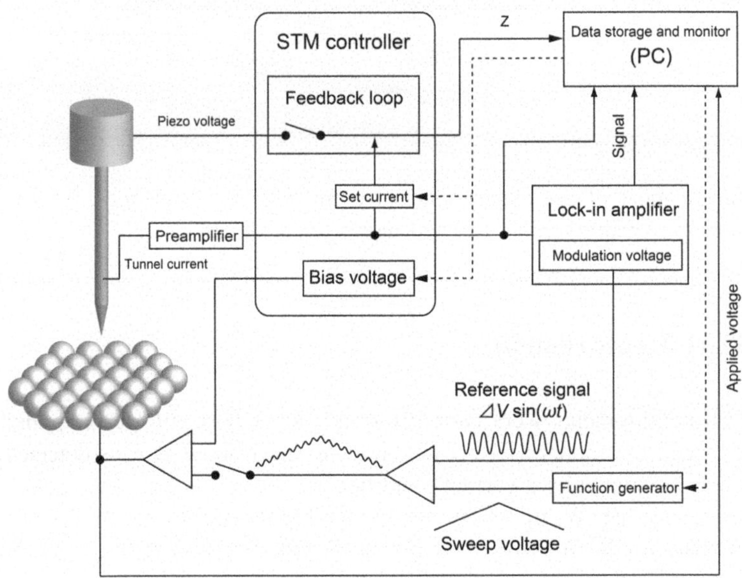

Fig. 3.4 Schematic diagram of the STM measurement system

interpretation of acquired images constant current mode is usually employed. Acquired STM images were processed using graphical analysis software. A quasi-3D image of surface corrugations is colored to improve its visualization. The color scale is shown with STM images.

dI/dV and d^2I/dV^2 spectra in an STM junction were measured using a lock-in amplifier. Figure 3.4 shows a schematic diagram of the measurement system. During the measurement an STM tip is fixed over a molecule with opening the feedback loop and the bias voltage superimposed by $\Delta V \sin(\omega t)$ is ramped. ΔV is the amplitude and ω is the frequency of the modulation. The tunneling current can be expanded as

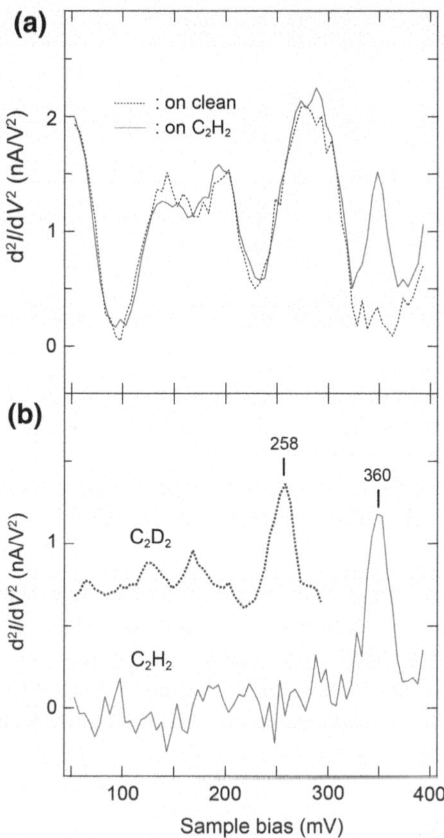

Fig. 3.5 **a** STM-IETS spectra for an isolated C_2H_2 molecule on Cu(110) (*red solid curve*) and clean surface (*black dashed curve*). The tip was fixed vertically to give $I_t = 50$ pA at $V_s = 0.1$ V. **b** The differential spectra show peaks at 350 and 258 mV for C_2H_2 (*red solid curve*) and C_2D_2 (*blue dashed curve*), respectively, which are assigned to the C–H(D) stretch mode of $\sim sp^3$ rehybridized acetylene molecule

$$
\begin{aligned}
I[V_0 + \Delta V \sin(\omega t)] &= I(V_0) + \frac{dI}{dV}\Delta V \sin(\omega t) + \frac{d^2I}{dV^2}\frac{(\Delta V)^2}{2}\sin^2(\omega t)\cdots \\
&= I(V_0) + \frac{dI}{dV}\Delta V \sin(\omega t) + \frac{d^2I}{dV^2}\frac{(\Delta V)^2}{4}(1 - \cos(2\omega t))\cdots
\end{aligned}
$$

thus dI/dV and d^2I/dV^2 signals can be extracted from ω and 2ω components, respectively. In order to verify our setup and stability, I measured the STM-IETS of an acetylene molecule on Cu(110). Since it was first studied on Cu(100) and the C–H(D) stretch mode was clearly observed [12], acetylene can be considered as a good referable molecule. Figure 3.5 shows the d^2I/dV^2 (IET) spectra of acetylene molecules adsorbed on Cu(110), where the C–H(D) stretch mode is also observed [13].

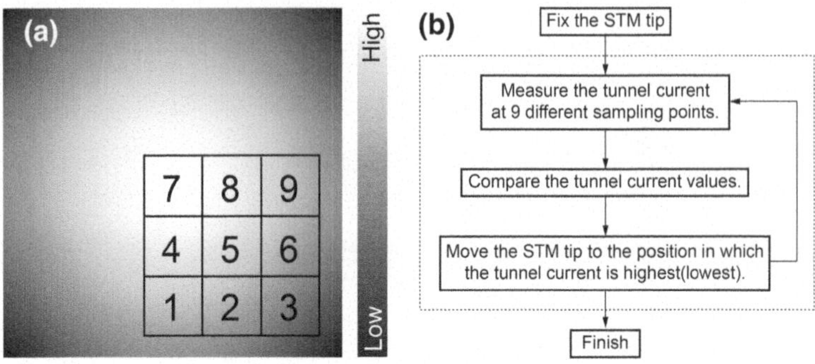

Fig. 3.6 Schematic illustration of our atom-tracking routine. An STM tip measures tunneling current at 9 different points as shown in **a** Then the measured current values are compared and the tip is moved to the position which gives the highest or lowest value. The cycle within the dashed line in **b** is repeated specified number of times

3.4 Time-Resolved Measurement of an STM

STM has an extremely high spatial resolution. The time resolution, however, is limited. It takes about a fraction of a second for a fast scanning STM to few minutes for a standard STM to obtain one image. These time scales are not enough to monitor atom/molecular motions and reactions. The poor time resolution can be improved up to few micro-seconds order using following techniques.

3.4.1 Atom-Tracking

Atom-tracking STM technique was introduced by Swartzentruber in 1996 [14], which relies on a tracking procedure [15]. In this method, an STM tip is first fixed over a preselected atom or molecule, and then the tip is laterally moved with a few angstroms to detect the local surface gradient within the x–y plane. The gradient is measured as a derivative of tunneling current using a lock-in amplifier. The STM tip is finally moved to a position of zero local gradients that is on top of the atom or molecule. It is possible to force the STM tip to stay over an atom or molecule during the atom tracking routine. Hence by recording the STM tip displacement from the original (preselected) position, we can trace the motion of atom or molecule.

I used an alternative routine for the atom tracking. The routine is quite simple as illustrated in Fig. 3.6. After fixing the STM tip at a preselected position (position 5), tunneling current or z position are measured at surrounding 8 positions (1–4 and 6–9) arranged in a square (Fig. 3.6a). Then the measured values of tunneling current (z position) at each point are compared and the STM tip is moved to a position of the maximum or minimum value. If we set the conditions to trace the maximum (minimum), the STM tip is moved to position 7 (3) in Fig. 3.6a. Repeating this loop (inside of dashed line in Fig. 3.6b) the tip is always forced to

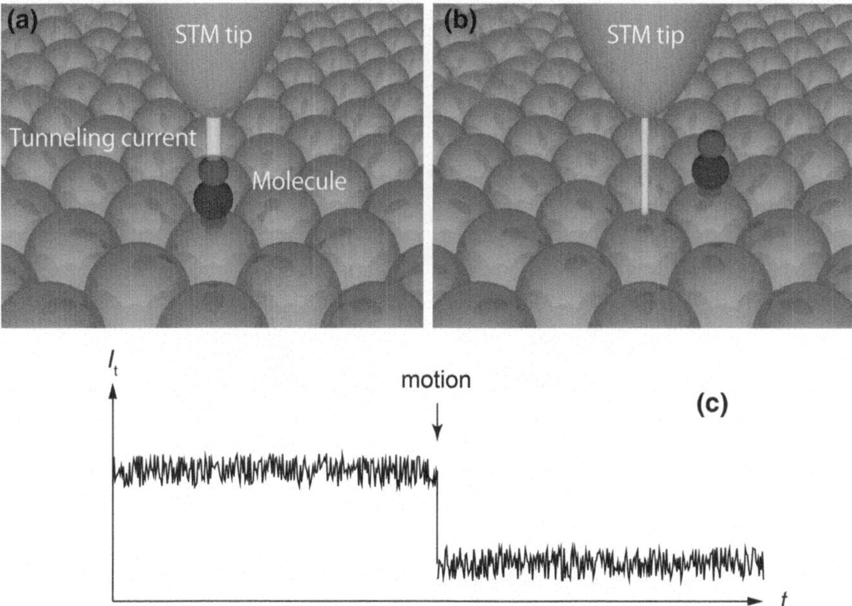

Fig. 3.7 Schematic illustration of open feedback measurement. **a** An STM tip is fixed over a molecule and then a feedback loop is opened. **b** A molecular motion under the tip results in a change in tunneling current. **c** The tunneling current is recorded during the measurement. A dynamical behavior can be observed as an abrupt change in the current trace

stay at the maximum or minimum position. The repetition rate of this loop is determined by the accumulating time for the data acquisition at each point, which is usually set to be 10–20 ms. The time resolution of this method is expected to be a few tenth of a second (practically $\sim$0.3 s).

3.4.2 Open Feedback Loop Measurement at a Single-Point

The poor time resolution of STM results from the feedback loop controlling the tip z position. If the feedback loop is opened with keeping the STM tip at a specific single-point in x–y, we can reach a higher time resolution. The open feedback loop measurement was first demonstrated by Lozano and Tringides in 1995 [16]. They studied surface diffusion by investigating the tunneling current as a function of time with opened the feedback loop. In the late 1990 s, Stipe et al. demonstrated the direct monitoring of a single molecule motion with open feedback loop measurement [17, 18]. Since an STM tip is fixed at a specific height in the open feedback loop measurement, dynamical phenomena under the tip can be directly observed as a change in tunneling current (Fig. 3.7). Molecular motions or

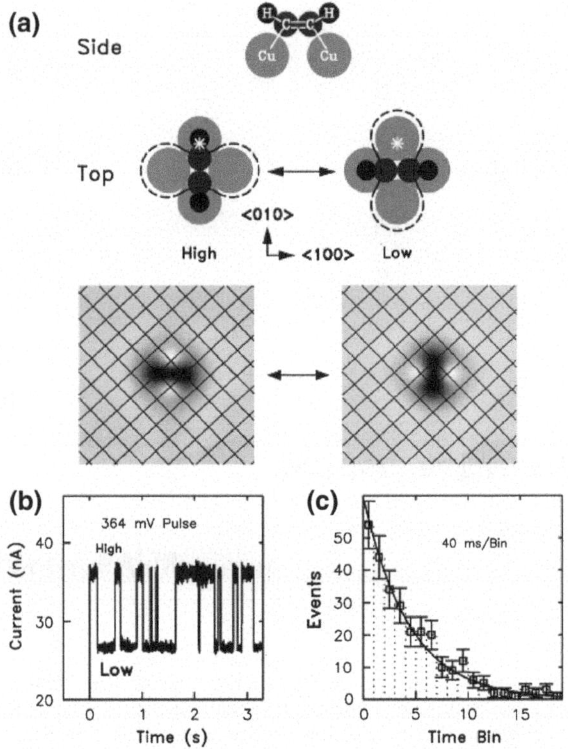

Fig. 3.8 **a** Schematic drawing of acetylene on a Cu(100) surface showing side and top views of the molecular adsorption site and orientations consistent with the STM images. The dashed line represents the outline of the dumbbell shaped depression in the images. The asterisk shows the off-center position of the tip over the molecule. In the corresponding STM images, the square lattice represents the Cu atoms of the substrate and was taken from an atomic resolution image of the same area with the C_2H_2 molecule adsorbed on the tip. The distance between Cu atoms is 2.55 Å. The images were scanned at a tunneling current of 10 nA and a sample bias of 100 mV. **b** Current during a 364 mV voltage pulse over a C_2H_2 molecule initially in the high current orientation; tip remains fixed 1.5 Å off center. Each jump in the current indicates the moment of rotation of the molecule. **c** Distribution of the times the molecule spent in the highcurrent orientation with a fit to an exponential decay. The fitted time constant is 184 ms. The analysis includes a total of 278 time intervals. Reprinted with permission from Ref. [18]. Copyright 1998, American Physical Society

reactions result in an abrupt change in a current trace. For instance, the rotation of a single molecule at the binding site (Fig. 3.8a) can be observed as a telegraph noise signal (Fig. 3.8b). Tunneling current fluctuates between the high and low current states corresponding to the two molecular orientations under the STM tip. From this measurement the time spent at the each state can be obtained and the rotation rate is determined by the statistical analysis of the distribution of the time intervals of the rotation events. The histogram of a time spent is fitted to the exponential decay function, $N = N_0\exp(-Rt)$, where N is the event number, N_0 is pre-exponential factor, R is the rate of the event and t is the time.

The open feedback loop measurement enables us to trace the dynamical behavior of adsorbates with time resolution up to a few hundreds micro-second. More recently, Loth and his co-workers at IBM Almaden research center achieved a few nano-seconds resolution by applying the electrical based pump-probe technique [19]. Using this technique they directly measured the lifetime of spin excitation of single magnetic atom. This technique would provide us an opportunity to capture very fast dynamics, such as a lifetime of molecular vibrations, if the resolution will be further improved.

References

1. A. Roth, *Vacuum Technology*, 2nd edn. (North Holland, Amsterdam, 1982)
2. N. Harris, *Modern Vacuum Practice* (McGraw-Hill, New York, 1989)
3. J.F. O'Hanlon, *A Users Guide to Vacuum Technology*, 2nd edn. (John Wiley, New York, 1990)
4. T. Yamada et al., Phys. Rev. Lett. **96**, 036105 (2006)
5. Carrasco et al., Nature Mater. **8**, 427 (2009)
6. T. Schiros et al., Chem. Phys. Lett. **429**, 415 (2006)
7. C. Ammon et al., Chem. Phys. Lett. **377**, 163 (2003)
8. J. Lee et al., J. Phys. Chem. C **112**, 17672 (2008)
9. M. Forster et al., Phys. Rev. Lett. **106**, 046103 (2011)
10. H.-W. Fink, Ibm J Res Dev **30**, 460 (1986)
11. S.C. Wang, T.T. Tsong, Phys. Rev. B **26**, 6470 (1982)
12. B.C. Stipe, M.A. Rezaei, W. Ho, Science **280**, 1732 (1998)
13. T. Kumagai, S. Hatta, H. Okuyama, T. Aruga, J. Chem. Phys. **126**, 234708 (2007)
14. B.S. Swartzentruber, Phys. Rev. Lett. **76**, 459 (1996)
15. D.W. Pohl, R. Möller, Rev. Sci. Instrum. **59**, 840 (1988)
16. M.L. Lozano, M.C. Tringides, Europhys. Lett. **30**, 537 (1995)
17. B.C. Stipe, M.A. Rezaei, W. Ho, Science **279**, 1907 (1998)
18. B.C. Stipe, M.A. Rezaei, W. Ho, Phys. Rev. Lett. **81**, 1263 (1998)
19. S. Loth, M. Etzkorn, C.P. Lutz, D.M. Eigler, A.J. Heinrich, Science **329**, 1628 (2010)

Chapter 4
Water Monomer: Structure and Diffusion of a Single Water Molecule

Abstract I describe the structure and diffusion of a single water molecule isolated on a Cu(110) surface in this chapter. The monomeric adsorption of water molecules can be realized at low temperature (<20 K). The adsorption structure is determined with the help of DFT calculations. A water molecule chemisorbs on a top of Cu atoms and thermally diffuses on the surface along the atomic row of Cu(110) even at 6 K. Using the time-resolved measurement of the STM, the diffusion rate is determined to be 0.13 s^{-1} that corresponds to the barrier of $\sim$15 meV. The diffusion is also induced by the vibrational excitation of water molecule via the inelastic electron tunneling process. I also show the controlled manipulation of individual water molecules. A water molecule can be moved along the atomic row direction by the lateral manipulation of STM.

Keywords Water monomer · Adsorption geometry · Surface diffusion · Molecule manipulation

4.1 Introduction

The adsorption of an isolated water molecule on metal surfaces provides a fundamental understanding about a bare interaction between a water molecule and metal surfaces. The question is simple as follow; where and how a water molecule adsorbs on a metal surface. In contrast with this simple question, the experimental characterization of an isolated water molecule is extremely difficult due to a facile clustering of molecules. In order to minimize the clustering, it is essential to perform experiments at extremely low coverages of water and at low temperatures. Monomeric water is not observed on most of noble metal surfaces even at liquid nitrogen temperature and it becomes possible below 20 K. On the basis of the molecular orbital of a free water molecule, it is expected that the interaction of water with a metal surface would be via the non-bonding, or lone pair electrons on the

T. Kumagai, *Visualization of Hydrogen-Bond Dynamics*, Springer Theses,
DOI: 10.1007/978-4-431-54156-1_4,

O atom. This geometry was examined using electron stimulated desorption ion angular distributions (ESDIAD) by Madey and Yates [1]. The monomeric adsorption on Cu(100) and Pd(100) surfaces was characterized using high-resolution electron energy loss spectroscopy (HREELS) at $\sim$10 K [2]. The HREELS measurement suggests a water molecule is adsorbed on the surface via the O atom with its molecular plane lying nearly parallel to metal surfaces. The parallel geometry to the surface was also predicted in a water monomer on the Ru(0001) surface at 20 K using infrared absorption spectroscopy [3, 4]. Although the experimental literature of the monomeric water adsorption is still limited, theoretical calculations predicted the adsorption structure on several kinds of noble metal surfaces [5–11], where the nearly parallel geometry is also predicted. It was also found that there is little distortion of the monomer upon adsorption, with only a slight elongation of the O–H bonds and opening of the HOH angle.

Diffusion of individual water molecules can be considered as an elemental process on metal surfaces and it directly reflects strength of the interaction between a water molecule and metal surfaces. Vibrational spectroscopies have been employed to investigate the diffusion and clustering processes, since the H-bond formation has significant impact on the O–H stretch vibrational energy [12]. However it is hard to know the local behavior of individual molecules by such macroscopic methods. The development of a low temperature STM has paved a novel and unique way to investigate the local structure and dynamical behavior of a water monomer and small clusters on metal surfaces. For instance, in 2002 Mitsui et al. directly observed the diffusion and aggregation process of water molecules adsorbed on a Pd(111) surface at 40 K [13]. They presented the real-space and time observation of the diffusion of a water monomer and small clusters. Furthermore, the diffusion rate of each species was quantified using the atom-tracking technique. In 2001, Lauhou and Ho reported a direct observation of monomeric water on a Cu(100) surface and investigated the interaction between a water molecule and various co-adsorbates [14]. They observed the thermal diffusion of a water molecule and the hydrogen abstraction reaction from a water molecule to another chemical species at 9 K. In 2008, Motobayashi et al. observed a monomeric water on the Pt(111) surface at 4.7 K [15], where the thermal diffusion is not observed due to the relatively strong interaction between a water molecule and Pt surface.

4.2 Results and Discussions

4.2.1 Adsorption Structure of a Water Monomer on Cu(110)

Figure 4.1a shows a typical STM image of water monomers on Cu(110). A monomer is characterized by a round protrusion. This appearance is in common with the previous results observed on Pd(111), Cu(100), and Pt(111) surfaces [13–15]. The adsorption site is important to understand the interaction of water with metal surfaces. Figure 4.2 illustrates possible adsorption sites on a Cu(110) surface. There are four high-symmetry adsorption sites; on-top(a), short-bridge(b), long-bridge(c) and hollow sites(d). An unambiguous way to determine the adsorption site with STM is observing an atomic arrangement of the substrate and molecules at the same time. It is, however, extremely difficult due to the small corrugation of Cu surface. Therefore I employed an alternative way as follows; a maker molecule whose adsorption site is well-established is co-adsorbed with water, and the water location can be determined by comparing the relative position between them. A carbon monoxide (CO) molecule is well-established to be adsorbed on a top site [16] on Cu(110). In addition, CO molecules can be easily identified by STM-IETS where the hindered rotation is observed. According to the relative position of water molecules to CO, the adsorption site is determined to be a top site (inset of Fig. 4.1a). The adsorption structure of water was determined by DFT calculations.[1] Figure 4.1b shows the optimized structure and the on-top configuration is very consistent with the experiment. The adsorption energy is 0.34 eV and the Cu–O bond length is 2.21 Å. The water adsorption is accompanied by a charge transfer from the occupied lone pair orbital of the molecule ($1b_1$ orbital) to unoccupied metal *d*-orbitals, consequently, a water molecule prefers to occupy the electron deficient on-top adsorption sites. It is noted that the DFT calculations indicated the water molecule shows a slight displacement from the exact top site and the displacement from the center of a substrate atom is ~0.5 Å. Ren and Meng also predicted the similar displacement by 0.5 Å [11], and Tang and chen reported the displacement of 0.88 Å [10]. The center of protrusion in the STM image, however, is exactly positioned on the on-top site. The calculated barrier to the azimuthal rotation is less than 0.02 eV, implying almost free rotation of a water molecule around the on-top site, resulting in the featureless round protrusion. It might be suggested that the facile rotation of a water monomer arises from its zero-point motion.

4.2.2 *Diffusion of a Water Monomer on Cu(110)*

It was found that water molecules thermally diffuse along the atomic row even at 6 K on Cu(110). The fractional image of the upper right monomer in Fig. 4.1a indicates that the molecule hops to the adjacent on-top site when the STM tip scanned over it. Not only the thermal activation but also the tip-induced diffusion must be taken into account in STM measurement. The contribution of the latter depends on tunneling conditions, i.e., the gap distance and the applied bias voltage. To elucidate the diffusion mechanism, I investigated the diffusion rate using the time-resolved measurement of the STM. Two different techniques were employed to determine the diffusion rate. At low bias voltages, the atom tracking technique

[1] DFT calculations were performed using the STATE code [Y. Morikawa et al. Phys. Rev. B 69, 041403 (2004).]. In the calculation the water molecules is put on one side of five-layer Cu slabs and 2×3 surface unit cell with 16 *k* point sampling.

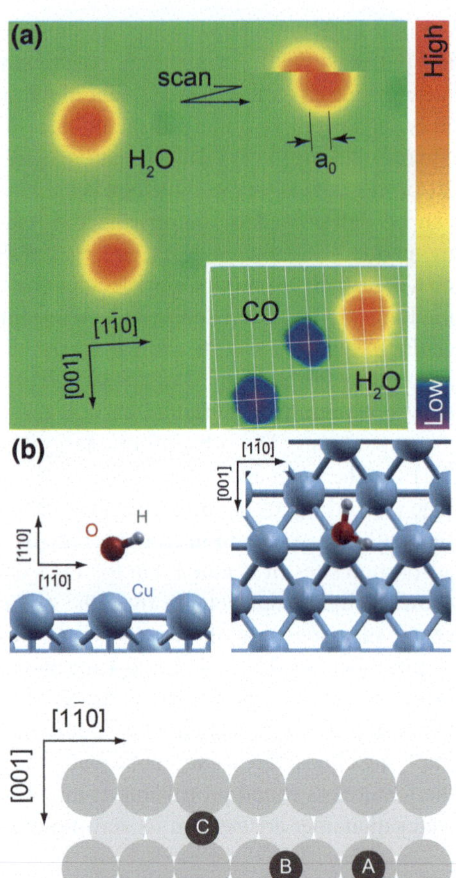

Fig. 4.1 **a** Typical STM image (53 × 53 Å^2) of single water molecules on a Cu(110) surface. The image was acquired at V_s = 24 mV and I_t = 0.5 nA. A water molecule appears as a round protrusion with the apparent height of 0.6 Å. The tip was scanned horizontally from top to bottom. The inset image shows relative position of a water molecule to co-adsorbed CO molecules. The *white grid lines* represent the lattice of Cu(110). The image was acquired at V_s = 24 mV and I_t = 0.2 nA. **b** Side (*left*) and top (*right*) view of the optimized structure for a water molecule on a Cu(110) surface

Fig. 4.2 Schematic illustration of the possible high symmetry adsorption sites on a Cu(110) surface. **a** on-top site, **b** short-bridge site, **c** long-bridge site and **d** hollow site

was used, in which the STM tip traced the molecular motion in the $x-y$ plane and its trajectory was recorded [17, 18]. Figure 4.3a shows a typical trajectory of a water monomer measured at V_s = 24 mV and I_t = 0.5 nA. The displacements from the original position along the [1$\bar{1}$0] (solid red line) and [001] (dashed blue line) directions are plotted as a function of the measurement time. The trajectory clearly indicates a water molecule diffuses exclusively along the atomic row direction. The distribution of the time intervals between hopping events is plotted in Fig. 4.3b. The distribution is fitted by an exponential decay function, $N = N_0\exp(-Rt)$, where R is the diffusion rate. Consequently, R is determined to be 0.13 s^{-1} at V_s = 24 mV and I_t = 0.5 nA. The inset of Fig. 4.3b shows the current dependence of the diffusion rate at V_s = 24 mV. The rate is almost constant (~0.13 s^{-1}) at less than I_t = 0.5 nA, suggesting the tip effect is negligible and the diffusion is induced via merely thermal activation in this voltage region. Given the

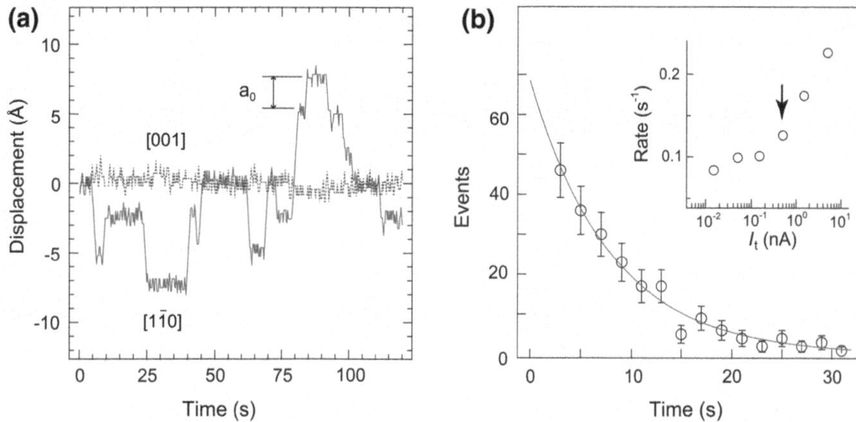

Fig. 4.3 **a** Typical trajectory of the STM tip when it tracks a water molecule on a Cu(110) surface at $V_s = 24$ mV and $I_t = 0.5$ nA. The *solid* (*red*) and *dashed* (*blue*) *lines* indicate the displacements along the [1$\bar{1}$0] and [001] directions, respectively. **b** Distribution of the time intervals between the hopping events, where the tunneling conditions are the same as (**a**). Fitting to an exponential function (*solid red line*) provides the diffusion rate of 0.13 s^{-1}. The inset shows the rate as a function of tunneling current at $V_s = 24$ mV measured for H_2O

pre-exponential factor of $\nu_0 = 10^{13}\ s^{-1}$, the barrier of the hopping is estimated to be 15 meV according to $R_0 = \nu_0 \exp(-E_b/k_B T)$, where E_b is the barrier height, k_B is the Boltzmann constant, and T is the temperature (6 K). This barrier is comparable with the value expected for a water molecule on Cu(100) (~20 meV) [19]. On the other hand, the diffusion rate shows an increase above $I_t \sim 0.5$ nA, indicating the tip-molecule interaction affects the diffusion process due to the reduced gap (tip-molecule) distance. The diffusion rate also increases even when the bias voltage is reduced to ~5 mV (the reduction of the bias voltage also reduces the gap distance). At present, the underlying mechanism is not clear, but it could be ascribed to the proximity between the tip and molecule, which presumably perturbs the potential energy surface of the hopping.

The STM-induced diffusion is also observed at higher bias voltages. Since the diffusion rate becomes too fast to follow up with the atom tracking technique at a higher bias region, I employed the open feedback measurement to observe the hopping event. Before opening the feedback loop the tip was fixed over a water molecule at low bias conditions, and then a voltage pulse for few seconds was applied while the current signal was recorded. A typical current-time plot is shown in the inset of Fig. 4.4a (measured at $V_s = 54$ mV). The abrupt drops in the current trace correspond to the moments of the hopping event. The first and second drops indicate the moments of hopping to the first- and second-neighbor site, respectively. The hopping rate is measured as a function of the applied bias voltage (Fig. 4.4a). The tip height is fixed to give $I_t = 0.012 \pm 0.003$ nA at $V_s = 24$ mV in which the tip proximity effect is negligible (inset of Fig. 4.3b). Figure 4.4a shows the voltage dependence of the hopping rate for an H_2O and D_2O monomer.

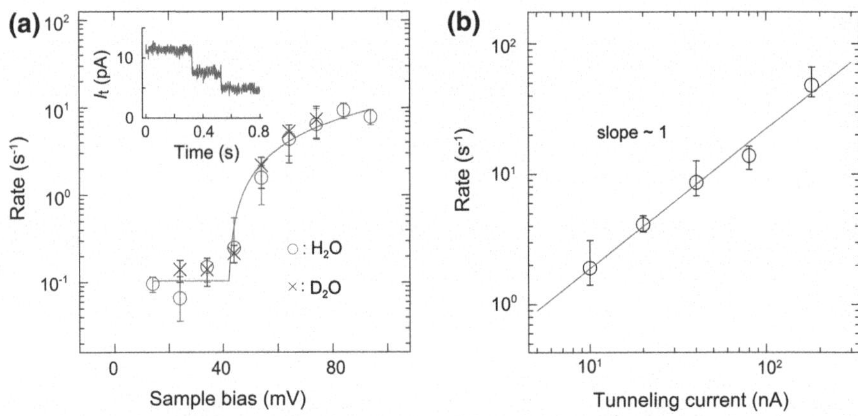

Fig. 4.4 **a** The diffusion rates as a function of bias voltage for H_2O (*red circles*) and D_2O (*blue crosses*). The tip height was adjusted to give $I_t = 12 \pm 3$ pA during the voltage pulses. The inset shows typical tunneling current taken with the tip fixed over the molecule at $V_s = 54$ mV. **b** The rate of induced motion as a function of I_t at $V_s = 54$ mV gives a slope of 1.0 ± 0.1 in a logarithmic scale

While the rate is constant below ~40 mV, the rate increases above ~40 mV without any isotope effects. The same behavior is observed at a negative bias voltage. The current dependence of the rate at $V_s = 54$ mV is also investigated (Fig. 4.4b). In Fig. 4.4b the contribution of the thermally-activated hopping ($R_0 = 0.13\ s^{-1}$) is subtracted to evaluate the rate purely induced by STM. If a reaction or motion of adsorbates is induced by electrons, its rate is proportional to the power low: $R \propto I^N$. Here power low index N corresponds to the reaction order, thus, the number of electron required to induce a reaction. The rate shows a linear dependence ($N \sim 1$) onto tunneling current at $V_s = 54$ mV, indicating that the hopping is induced via a single-electron process. The linear dependence also excludes the effect of electric field. These observations indicate increase of the hopping rate stems from the vibrational excitation of a water molecule via inelastic electron tunneling (IET) process.

In the IET framework, the rate of a molecular motion or reaction, R, is expected to be proportional to the inelastic current, $I_{\text{inelastic}}$, above a specific threshold [20]

$$R = KI_{inelastic}$$

and $I_{\text{inelastic}}$ is proportional to bias voltages

$$I_{inelastic} = K'V_s$$

Accordingly, the voltage dependence in Fig. 4.4(a) can be fitted to a linear function and constant above and below the threshold, respectively (red curve)

$$R = \begin{cases} R_0(V_s \leq V_{threshold}) \\ K'V_s(V_s \geq V_{threshold}) \end{cases}$$

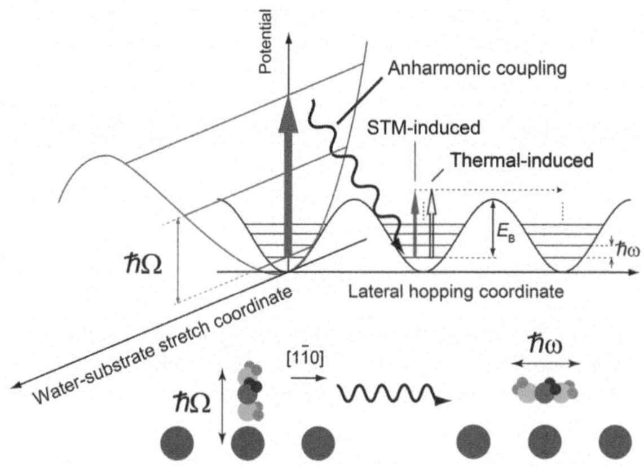

Fig. 4.5 Schematic illustration of the diffusion mechanism of a water molecule on a Cu(110) surface

where R_0 is the thermal diffusion rate (0.13 s^{-1}). The solid line in Fig. 4.4a shows the fitting curve and the threshold voltage is determined to be 43 mV for both of H_2O and D_2O. It is noted that the threshold voltage is much larger than the hopping barrier of 15 meV estimated from the thermal diffusion rate, indicating the energy of tunneling electron is not directly transferred into the hopping coordinate that is the hindered translation mode along the $[1\bar{1}0]$ direction. Although there is no available experimental data in literatures, the energy of the water-substrate (H_2O–Cu) stretch mode was calculated to be 26 meV by the normal-mode analysis of DFT calculations. Therefore, I tentatively assign the threshold energy to the excitation of the H_2O–Cu stretch or its overtone. Libration modes, such as wagging mode, were ruled out because of the absence of isotope effect in the threshold. On a Pd(111) surface, the hopping motion does not occur until the scissors mode excitation ($\sim$200 meV) [21]. This results from a difference in the barrier of hopping, which is 15 and 126 meV for Cu(110) and Pd(111) surfaces, respectively. Figure 4.5 illustrates the STM-induced diffusion process of water molecule on Cu(110). In addition to the thermal-induced process, the water-substrate stretching is excited by a single-electron process and relaxed into the hindered translation mode via anharmonic coupling, leading the molecule to overcome the hopping barrier.

4.2.3 Manipulation of a Water Monomer on Cu(110)

Here I show the STM manipulation of a water molecule. The manipulation of a single atom and molecule is an important application of a tip-adsorbate interaction of STM. In this way, adsorbates can be moved from one place to the other at the atomic-scale precision. As mentioned in the previous chapter, "atom manipulation" is first demonstrated by Eigler and Schweitzer. After their pioneering effort, various types of the STM manipulation have been developed [22] such as pulling [23],

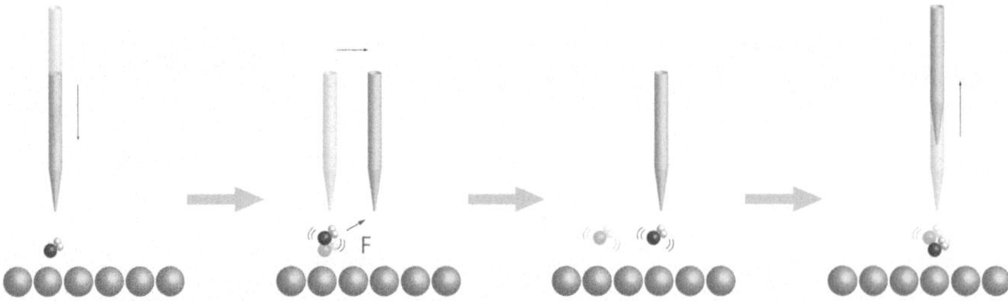

Fig. 4.6 Schematic illustration of water manipulation procedure. First the tip is brought close to a water molecule, and then the tip is laterally moved along the atomic row direction. A molecule follows the tip during moving the tip. Finally, the tip is retracted from the molecule

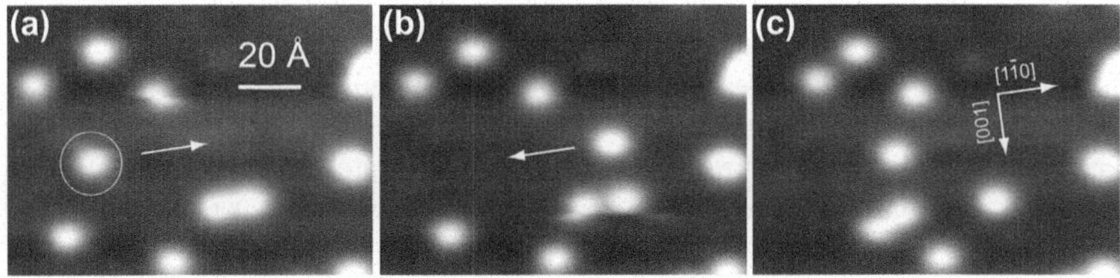

Fig. 4.7 Sequential STM images during the water manipulation. The molecule marked by the *white circle* is laterally moved to *right* **a–b** and *left* **b–c**. These images were obtained at $V_s = 24$ mV and $I_t = 0.5$ nA (84×117 Å^2)

pushing [24], flicking [25], and catching/releasing [26, 27] with STM. STM manipulation is considered to be an ultimate bottom-up process and used for assembling model systems of molecular devices. For instance, Heinrich and his co-workers at IBM demonstrated the assembly and operation of a molecular circuit using a precise positioning of CO molecules on Cu(111) [27].

A water molecule on Cu(110) can be moved along the atomic row direction in the following manner (described in Fig. 4.6); first the bias voltage is reduced to ~5 mV while the current is increased to ~10 nA over a water molecule (this procedure reduces the tip-molecule distance), and a tip is laterally moved at 1 Å/s along the atomic row direction with the feedback loop on. Then a water molecule follows the tip. Finally the tip is retracted to normal tunneling conditions. The above parameters are just one of the typical values and the suitable parameters for the stable manipulation depend on the tip apex conditions. On the other hand, the migration to the next atomic row cannot be induced even if a tip is moved along [001] direction, probably because the potential barrier over a trough between the atomic rows of Cu(110) is relatively high. The sequential images of the water manipulation are shown in Fig. 4.7. At first, the water molecule (indicated by white arrow) was manipulated in the $[1\bar{1}0]$ direction (Fig. 4.7a, b), and then it was moved backward (Fig. 4.7b, c).

4.3 Summary

A single water molecule on Cu(110) was imaged and characterized by STM. A water molecule chemically adsorbs on top of a Cu atom below 20 K. The adsorption structure was determined by the help of DFT calculations, suggesting the water molecule adsorbs the on-top site with a slight displacement and a nearly parallel configuration to the surface. The thermal diffusion along the atomic row takes place even at 6 K with the rate of 0.13 s^{-1}. Furthermore, the voltage and current dependence of the diffusion rate showed the STM-induced process was involved at the bias voltages above $\sim$40 mV, which results from the vibrational excitation of the water-substrate stretching via inelastic electron tunneling process. The mechanism is rationalized by anharmonic coupling between the water-substrate stretching and the hindered translational mode correlating the hopping coordinate. It was also found that water molecules can be manipulated in a controlled fashion along the atomic row.

References

1. T.E. Madey, J.T. Yates Jr, Chem. Phys. Lett. **51**, 77 (1977)
2. S. Andersson, C. Nyberg, C.G. Tengstål, Chem. Phys. Lett. **104**, 305 (1984)
3. M. Nakamura, M. Ito, Chem. Phys. Lett. **325**, 293 (2000)
4. S. Yamamoto, A. Beniya, K. Mukai, Y. Yamashita, J. Yoshinobu, J. Phys. Chem. **109**, 5816 (2005)
5. S. Meng, L.F. Xu, E.G. Wang, S. Gao, Phys. Rev. Lett. **89**, 176104 (2002)
6. S. Meng, E.G. Wang, S. Gao, Phys. Rev. B **69**, 195404 (2004)
7. A. Michaelides, V.A. Ranea, P.L. de Andres, D.A. King, Phys. Rev. Lett. **90**, 216102 (2003)
8. A. Michaelides, Appl. Phys. A **85**, 415 (2003)
9. V.A. Ranea, A. Michaelides, R. Ramírez, J.A. Vergés, P.L. de Andres, D.A. King, Phys. Rev. B **69**, 205411 (2004)
10. Q.-L. Tang, Z.-X. Chen, Surf. Sci. **601**, 954 (2007)
11. J. Ren, S. Meng, Phys. Rev. B **77**, 054110 (2008)
12. S. Yamamoto, A. Beniya, K. Mukai, Y. Yamashita, J. Yoshinobu, J. Phys. Chem. B **109**, 5816 (2005)
13. T. Mitsui, M.K. Rose, E. Fomin, D.F. Ogletree, M. Salmeron, Science **297**, 1850 (2002)
14. L.J. Lauhou, W. Ho, J. Phys. Chem. B **105**, 3987 (2001)
15. K. Motobayashi, C. Matsumoto, Y. Kim, M. Kawai, Surf. Sci. **602**, 3136 (2008)
16. M. Doering, J. Buisset, H.-P. Rust, B.G. Briner, A.M. Bradshaw, Faraday Discuss. **105**, 163 (1996)
17. B.S. Swartzentruber, Phys. Rev. Lett. **76**, 459 (1996)
18. M. Krueger, B. Borovsky, E. Ganz, Surf. Sci. **385**, 146 (1997)
19. M.A. Henderson, Surf. Sci. Rep. **46**, 1 (2002)
20. H. Ueba, Surf. Sci. **601**, 5212 (2007)
21. E. Fomin, M. Tatarkhanov, T. Mitsui, M. Rose, D.F. Ogletree, M. Salmeron, Surf. Sci. **600**, 542 (2006)
22. P. Avouris, Acc. Chem. Res. **28**, 95 (1995)
23. L. Bartels, G. Meyer, K.-H. Rieder, Phys. Rev. Lett. **79**, 697 (1997)

24. L. Bartels, G. Meyer, K.-H. Rieder, D. Velic, E. Knoesel, A. Hotzel, M. Wolf, G. Ertl, Phys. Rev. Lett. **80**, 2004 (1998)
25. M. Ohara, Y. Kim, M. Kawai, Phys. Rev. B **78**, 201405 (2008)
26. H.J. Lee, W. Ho, Science **286**, 1719 (1999)
27. A.J. Heinrich, C.P. Lutz, J.A. Gupta, D.M. Eigler, Science **298**, 1381 (2002)

Chapter 5
Water Dimer: Direct Observation of Hydrogen-Bond Exchange

Abstract I describe the direct observation of H-bond exchange reaction within a water dimer isolated on a Cu(110) surface in this chapter. A water dimer consisted of H-bond donor and acceptor molecule can be produced by the association of water monomers. The STM image of a dimer is characterized by an incessant fluctuation between two states. Combined with DFT calculations it is found that the fluctuation corresponds to the H-bond donor–acceptor interchange where the role of each molecule is exchanged within a dimer. To elucidate the mechanism the interchange rate is investigated using the time-resolved measurement of STM. The voltage and current dependence of the rate unveil the interchange is not induced by STM, thus inherent in a dimer, at low bias voltages ($V_s < 40$ mV). Furthermore, it is found that the rate of $(H_2O)_2$ is ~60 times larger than that of $(D_2O)_2$ in the low voltage region, indicating that the interchange process includes quantum tunneling. The voltage dependence also reveals the interchange rate is increased by the excitation of the intermolecular vibration mode of a dimer through the vibrationally assisted tunneling process.

Keywords Water dimer · Hydrogen-bond rearrangements · Quantum tunneling · Vibrationally assisted tunneling

5.1 Introduction

Water dimer has been an enthusiastic research object as a prototype of much more complex H-bonding systems and called “a theoretical Guinea pig”. A dimer is the smallest water cluster composed of H-bond donor and acceptor molecules (Fig. 5.1) and regarded as the simplest model of H-bond dynamics. The measured distance between molecules was 2.98 Å [1–4] that is significantly longer than that in both liquid water and regular ice (about 2.85 and 2.74 Å, respectively). The

T. Kumagai, *Visualization of Hydrogen-Bond Dynamics*, Springer Theses,
DOI: 10.1007/978-4-431-54156-1_5,

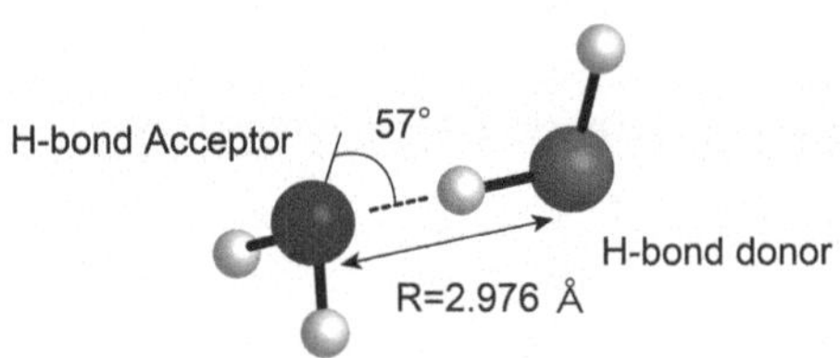

Fig. 5.1 The equilibrium structure of water dimer determined by calculations. The H bond deviates 2.3° from linearity, the O–O distance is 2.952 Å, and the bond strength is 3.40 kcal/mol

shortening of the R(O···O) distance in condensed phase represents the existence of a stronger H-bonded interaction and the cooperative nature of H bond.

The unambiguous insights into the structure and dynamics of a gas phase dimer were provided by the high-resolution vibration–rotation spectroscopy [1–8]. The pioneering work by Dyke and co-workers revealed that the H-bond exchange occurs via quantum tunneling in a water dimer [1–4] and ab initio calculations predicted three distinct low barrier tunneling pathways as shown in Fig. 5.2 [9]. First, in "acceptor switching" (Fig. 5.2a), having the lowest barrier of ~20 meV, the acceptor molecule is inverted where the H bond remains intact during the process. The second lowest barrier process is "donor–acceptor interchange" (Fig. 5.2b) where the role of each molecule is exchanged via the concerted rotation and the barrier is ~26 meV. Finally, "bification" is the highest barrier process having ~49 meV (Fig. 5.2c), where the bound and free hydrogen on the donor are exchanged. Here I focus on the process of donor–acceptor interchange. Figure 5.3 illustrates the potential energy surface of the interchange. The interchange proceeds through the concerted rotation of each molecule, wherein the transition state has C_{2v} symmetry. As described in the Chap. 2, quantum tunneling manifests the splitting of the energy level and its width represents the tunneling rate. This energy splitting was directly observed in the vibration–rotation-tunneling spectra. The tunneling rate for the interchange was estimated to be $\sim 10^9\ s^{-1}$ for $(H_2O)_2$ [5]. Moreover, the effect of vibrational excitation on the interchange tunneling dynamics was also investigated [10] and the tunneling rate was found to be affected drastically by the excitation of the intermolecular vibration mode that was argued to correlate with the interchange reaction coordinate.

In general, it is suspicious if we could directly apply the knowledge of H-bond dynamics of gas-phase species to that of adsorbate systems in which strong environmental effects must be included. However, the donor–acceptor interchange in a gas phase dimer was recently invoked to explain the unique mobility of a water dimer on a Pd(111) surface. In 2002, Mitsui et al. observed the anomalous diffusion rate of a water dimer on Pd(111) at 40 K, where the rate of the dimer was much faster than that of a monomer and other clusters [11]. Specifically, the diffusion of a water dimer was found to be larger than that of monomer and trimer (tetramer) by more than four and two orders of magnitude, respectively. They claimed the rapid diffusion is associated with the mismatch between the O–O distance of a dimer (2.95 Å in the gas phase) and the Pd lattice distance at the surface (2.75 Å). After Mitsui's report Ranea et al.

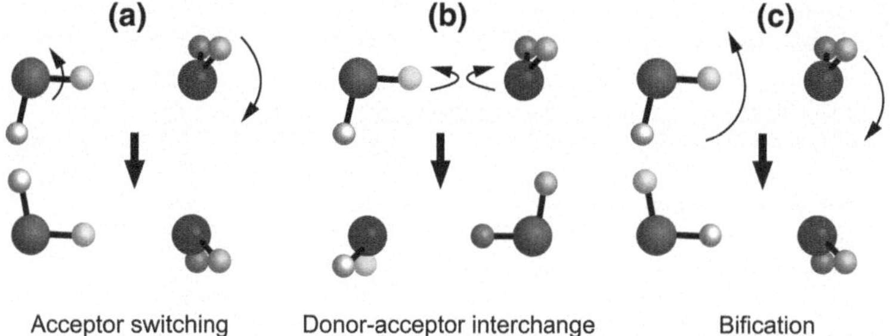

Fig. 5.2 Three distinct low barrier tunneling pathways that rearrange the H bond within a water dimer. **a** Acceptor switching that has the lowest barrier of all tunneling pathways, where the tunneling pathway exchanges the two H in the H-bond acceptor molecule and has been determined to begin with a flip of the acceptor molecule followed by a rotation of the donor molecule around its donating O–H bond, and completed by a 180° rotation of the complex about the O–O bond. **b** Donor–acceptor interchange in which the roles of the H-bond donor and acceptor molecules are exchanged. Several possible pathways exist for this interchange, the lowest barrier path being the geared interchange motion. This pathway proceeds via concerted rotation of the donor and acceptor molecules and through a *trans* transition state structure as shown in Fig. 5.3. **c** The bifurcation tunneling motion, where the H bond donor exchanges its protons, consists of the simultaneous in plane librational motion of the donor with the flip of the acceptor monomer. This is the highest barrier tunneling pathway

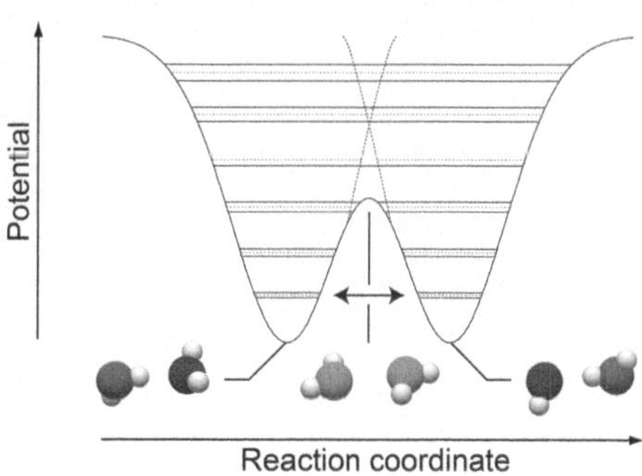

Fig. 5.3 Schematic illustration of energy diagram of the donor–acceptor interchange. The energy levels split into doublet due to tunneling

theoretically described the abnormal diffusion behavior [12]. They revealed the donor–acceptor interchange governed by quantum tunneling plays a crucial role in the diffusion process. Figure 5.4 shows the proposed process of a water dimer on Pd(111). The theoretical calculations predicted the dimer is bonded to the surface mainly via the oxygen atom of the donor molecule, and almost barrier-less azimuthal rotation of the acceptor around the donor (Fig. 5.4a, b. The diffusion proceeds through the wagging motion, which brings both water molecules to a similar height above the surface (Fig. 5.4b, c), and the subsequent donor–acceptor interchange (Fig. 5.4c–e). After the interchange the dimer restores its equilibrium configuration (Fig. 5.4e, f). The lowest diffusion barrier of this process is calculated to be 0.26 eV. This result does not rationalize the rapid diffusion of the dimer compared to monomer having the barrier of 0.19 eV [11]. However the inclusion of H-atom tunneling in the

Fig. 5.4 Mechanism for water dimer diffusion on Pd(111) (side and top view). Step (**a**) to (**b**) involves a nearly free rotation of the dimer; step (**b**) to (**c**) is the wagging motion of the dimer, which brings both water molecules to a similar height above the surface from where they can undergo donor–acceptor tunneling interchange (**c**)–(**e**). From step (**e**) to (**f**), the dimer restores its equilibrium geometry having translated one lattice spacing [compare (**a**) and (**f**)]. Reprinted with permission from Ref. [12]. Copyright 2004, American Physical Society

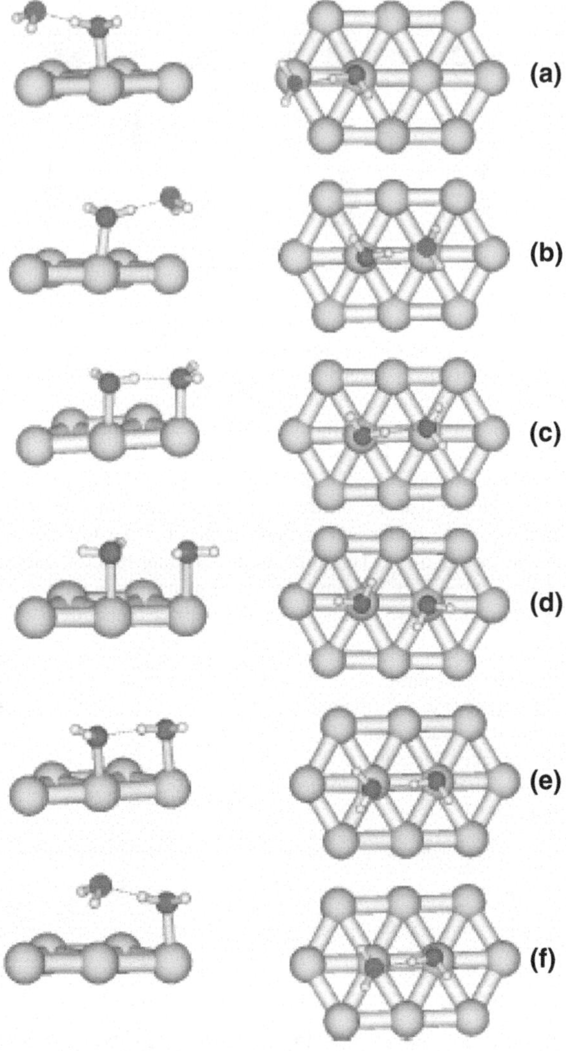

interchange process eventually leads to a considerably high diffusion rate of a dimer, which can explain the experimental findings. In summary, they presented the tunneling interchange has a significant impact on the diffusion of a water dimer on Pd(111) and concluded as follows; (i) For dimers a crossover from a quantum to a classical diffusion regime is expected at low temperatures. (ii) Diffusion of dimer should exhibit an isotope effect (a D_2O dimer should diffuse 10 times more slowly than an H_2O dimer in the tunneling regime) (iii) Dimers diffuse faster than monomers below a specific temperature. Although this finding shed light on the role of quantum dynamics of H-bond within small water clusters on surfaces, the direct observation of such process had never reported so far.

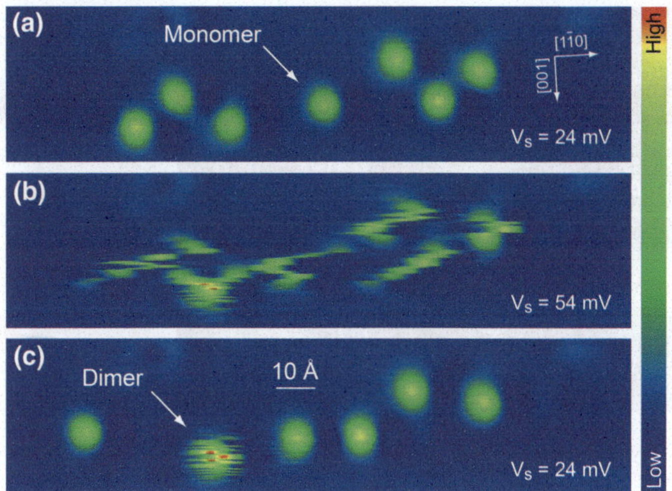

Fig. 5.5 Production of a water dimer on a Cu(110) surface **a** Constant current image of H_2O monomers. The image was acquired at $V_s = 24$ mV, $I_t = 0.5$ nA. **b** Scanning with $V_s = 54$ mV and $I_t = 0.5$ nA induces the hopping of monomer along [1$\bar{1}$0] direction. Two monomers encounter and yield a dimer. **c** The image after (**b**). The dimer is characterized by its fluctuating image. The apparent heights of the monomer and dimer are 0.6 and 0.8 Å, respectively. The size of these images is 40 × 160 Å^2

5.2 Results and Discussions

5.2.1 Production of a Water Dimer on Cu(110)

Figure 5.5 show the production of a water dimer. Round protrusions in Fig. 5.5a are water monomers. In order to associate water monomers, hopping along the atomic row direction is induced by scanning the surface at $V_s = 54$ mV (Fig. 5.5b). Then two monomers encounter with each other, and yield a dimer (Fig. 5.5c). A dimer is characterized by a continuously fluctuating image. A dimer was never dissociated into monomers spontaneously under the same conditions as its production, suggesting it is energetically preferred to isolated monomers due to H bond formation. The dissociation of dimer into monomers can be induced by applying a voltage pulse above 0.2 V over a dimer. This energy implies the H_2O scissors mode is associated with the dissociation mechanism.

Upon the deuterium (D) substitution, the STM image is drastically affected. Figure 5.6a shows the high resolution image simultaneously acquired for an $(H_2O)_2$ and $(D_2O)_2$. Since the STM tip is scanned horizontally from the top to bottom, the fluctuating image of the $(H_2O)_2$ corresponds to the motion along the atomic row direction. As discussed later in details, this fluctuation corresponds to the donor–acceptor interchange of a dimer. In contrast to the $(H_2O)_2$, the $(D_2O)_2$ is imaged stationary in Fig. 5.6a. Although the fluctuation takes place in a $(D_2O)_2$,

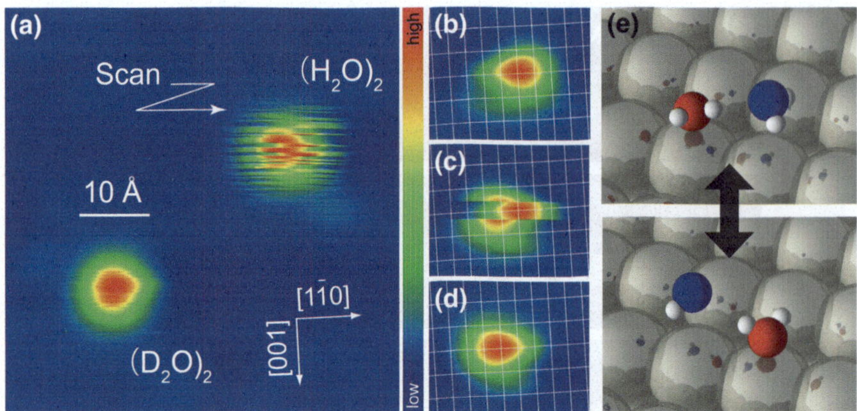

Fig. 5.6 **a** Typical STM image simultaneously acquired for an $(H_2O)_2$ and $(D_2O)_2$ (55 × 55 Å^2). **b** An enlarged image of a $(D_2O)_2$ (24 × 20 Å^2). The asymmetric egg shape reflects the donor–acceptor configuration of the water dimer along [1$\bar{1}$0]. The *white grid lines* represent the lattice of Cu(110). **c** The $(D_2O)_2$ flipped between the two orientations as the tip was scanned horizontally across it. **d** The counterpart of (**b**) in the same area. All images were acquired at $V_s = 24$ mV, and $I_t = 0.5$ nA for (**a**) and 0.3 nA for (**b**)–(**d**). **e** Schematic illustration of donor–acceptor interchange motion on a Cu(110) surface.

the rate is much lower than that of an $(H_2O)_2$. As a result, it is possible to observe the stationary image for a $(D_2O)_2$. A $(D_2O)_2$ is characterized by an asymmetric "egg" shape protrusion having the long axis along the [1$\bar{1}$0] direction. It is noted that the characteristic egg-shape can be also discerned for each envelope of the fluctuating image of an $(H_2O)_2$, indicating that both $(H_2O)_2$ and $(D_2O)_2$ have the same structure but a quite different dynamical property. Figure 5.6b–d show enlarged images of $(D_2O)_2$, which shows the details of the flip motion. The white grid lines represent the lattice of Cu(110). In Fig. 5.6c, the dimer underwent the flip motion as the STM tip was scanned over it. These images clearly indicate that the dimer flips between the two orientations.

5.2.2 Structure of a Water Dimer on Cu(110)

The structure of water dimer was determined by DFT calculations.[1] The optimized structure is shown in Fig. 5.7a, b. The dimer consists of an H-bond donor and acceptor molecule adsorbed on a top of Cu atoms and they align along the [1$\bar{1}$0]

[1] The calculations were based on the density functional theory using a planewave, pseudopotential code STATE [Y. Morikawa, Phys. Rev. B 51, 14802 (1995).]. Five-layer Cu slabs with surface were used in the calculation and water dimers arrayed in a 2 × 3 surface unit cell and a k-space sampling at 16 points. Adsorbates and two top Cu layers were relaxed.

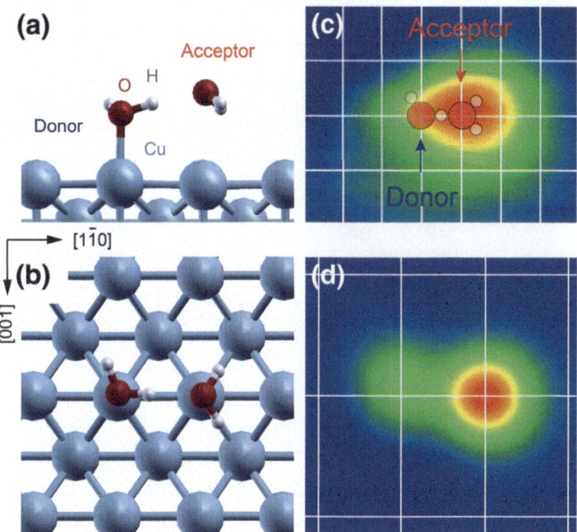

Fig. 5.7 **a** The optimized structure of a water dimer on a Cu(110) surface (side view). **b** The same as in (**a**) from top view. The Cu–O distances for the donor and acceptor molecules are 2.12 and 2.89 Å, respectively. The intermolecular O–O distance is 2.72 Å. **c** The experimental image for $(D_2O)_2$. **d** The calculated STM image corresponding to (**b**). The *white grid lines* represent the lattice of Cu(110)

direction. The donor (the lower-lying molecule) is directly bonded to a Cu atom, while the acceptor (the higher-lying molecule) interacts weakly with the substrate. This adsorption structure is in line with the previous calculations [13, 14]. The vertical displacement between the molecules was 0.67 Å, and the adsorption energy was 56 kJ/mol with that aligned the free dimer. In the calculations, the water dimer aligned along the [001] direction (bridging the atomic rows) had a comparable binding energy (58 kJ/mol) to the [1$\bar{1}$0] orientation. This configuration was, however, ruled out by the experimental result. The STM simulation[2] for the optimized structure is shown in Fig. 5.7d, which is characterized by the large and small protrusions over acceptor and donor molecules, respectively. The former reflects mainly $1b_1$ orbital (lone pairs) of the acceptor hybridized with substrate states. The simulation reproduces the experimental image qualitatively and the experimental image is assigned as shown in Fig. 5.7c. Now it is clear that the bi-stable fluctuation of a dimer results from the donor–acceptor interchange shown in Fig. 5.6e.

[2] The STM simulation was conducted based on the Tersoff-Hamann approach [J. Tersoff and D. R. Hamann, Phys. Rev. Lett. 50, 1998 (1983).] with a 3 × 4 unit cell and finer 64 k points. In the calculation of the simulation images, the sample bias voltage and the tip height were set at 24 mV and 0.5 nm, respectively. We confirmed that the qualitative features were not affected by the tip height. The molecular graphics were produced by the XCRYSDEN graphical package [A. Kokalj, Comput. Mater. Sci. 28, 155 (2003). Code available from http://www.xcrysden.org/]

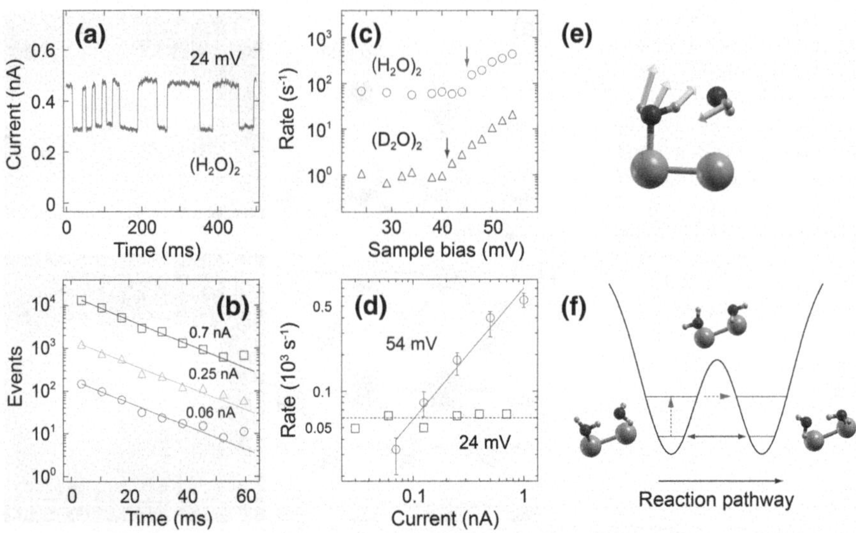

Fig. 5.8 **a** Typical current trace recorded with fixed the tip over an $(H_2O)_2$ at $V_s = 24$ mV and $I_t = 0.5$ nA. The tip was slightly displaced from the center of a dimer along $[1\bar{1}0]$ direction. The current jumps between the two states correspond to the moments of individual interchange events. **b** Distribution of the time intervals between the interchange events for an $(H_2O)_2$, which were obtained at three different tunneling current. The vertical axis is logarithmic scale. The distribution fitted to an exponential function in order to determine the interchange rate. **c** Voltage dependence of the interchange rate for an $(H_2O)_2$ (*open circles*) and $(D_2O)_2$ (*open triangles*). The tip height was adjusted to give 0.5 ± 0.1 nA for the high-current state during the measurement. The *arrows* indicate the threshold voltages at which the rate starts to increase. The thresholds were determined to be 45 ± 1 and 41 ± 1 mV for an $(H_2O)_2$ and $(D_2O)_2$, respectively. The tip height was varied so that the tunneling current was 0.06, 0.25, and 0.7 nA. The latter two data are displaced vertically for clarity. **d** Current dependence of the interchange rate at $V_s = 24$ mV (*open squares*) and $V_s = 54$ mV (*open circles*) for an $(H_2O)_2$ in a logarithmic scale. The latter rate indicates a linear dependence to the current. **e** A calculated normal mode for an $(H_2O)_2$ that couples with the interchange reaction. This mode involves the motions of donor-substrate stretch and acceptor rotation. **f** A schematic diagram of the potential energy along the interchange reaction pathway. The transition state is of C_{2v} symmetry with the Cu–O distances of 2.26 Å. In addition to the intrinsic tunneling between the ground states (*blue* or *dashed line double-ended arrow*), the interchange is mediated by the vibrational excitation (*red* or *solid line arrows*). The potential barrier is 23 kJ/mol (0.24 eV) while the vertical scale is not realistic

5.2.3 *Quantitative Analysis of the Donor–Acceptor Interchange of a Water Dimer on Cu(110)*

The interchange can be directly monitored by recording tunneling current over a water dimer. Figure 5.8a shows a typical current–time spectrum measured over an $(H_2O)_2$ at $V_s = 24$ mV. The feedback loop of the STM is turned off during the measurement. The tunneling current shows a bi-stable fluctuation corresponding to the interchange and current jumps correspond to the moments of individual interchange events. The distribution of the time intervals between the events was obtained at several different tunneling currents (gap resistance), and three of them are shown in Fig. 5.8b. Fitting the distribution to an exponential decay function provides the interchange rate. Here the slope represents the rate because the vertical axis in Fig. 5.8b is logarithmic scale. The slopes in Fig. 5.8b are almost independent on the three tunneling currents, indicating the tip effect to the interchange is negligible under the tunneling conditions employed. The interchange rate is determined to be $(6.0 \pm 0.6) \times 10\ s^{-1}$ for an $(H_2O)_2$, while it is $1.0 \pm 0.6\ s^{-1}$ for a $(D_2O)_2$. Thus the isotope ratio of the interchange rate is $\sim$60. Such a large isotope ratio indicates the process involves quantum tunneling.

It is assumed that the donor–acceptor interchange proceeds in a double-well potential as schematically shown in Fig. 5.8f. The DFT calculations predict the transition state of C_{2v} symmetry, and the barrier is calculated to be 23 kJ/mol on a Cu(110) surface while it is 21 kJ/mol on Pd(111). These barriers cannot be overcome via mere thermal process at experimental temperature, 6 K, being consistent with the interpretation that the interchange includes quantum tunneling. In the meantime, the interchange rate of a gas-phase $(H_2O)_2$ estimated from the tunneling splitting in vibration–rotation-tunneling spectra is $10^9\ s^{-1}$. The reduction of the rate by seven orders of magnitude partly results from the increase in the barrier height from 2.48 kJ/mol (gas-phase value) to 23 kJ/mol. In contrast to free dimers, the interchange requires substantial displacement of oxygen atoms for dimers adsorbed on surfaces. For a water dimer on Pd(111), the interchange process was divided into two, which were treated separately with classical and quantum models [12]. This hypothesis succeeded in rationalizing an anomalously higher mobility of water dimer than the monomer on Pd(111) at 40 K. For the dimer on Cu(110), however, the displacement of the acceptor molecule may be restricted due to relatively high barrier of 8 kJ/mol. Additionally, the azimuthal rotation of the acceptor around the donor, as observed on Pd(111), is inhibited on Cu(110) due to the anisotropic structure. I propose that tunneling proceeds along the optimal directions on the multidimensional potential energy surface with motions not only of hydrogen but also of oxygen atoms [15]. The multidimensional treatment of the tunneling process is essential for the quantitative description of the interchange motion.

I now turn to the voltage dependence of the interchange rate. Figure 5.8c shows the rate as a function of applied bias voltage. The rate shows no dependence and increase below and above 40 mV, respectively. The threshold voltage of the increase is determined to be 45 ± 1 (41 ± 1) mV for an H_2O (D_2O) dimer. The current dependence of the rate at $V_s = 54$ mV is also investigated (Fig. 5.8d). The rate is plotted as a function of tunneling current, where the intrinsic contribution ($60\ s^{-1}$) is subtracted from the rate to evaluate the interchange induced by tunneling electron. The interchange rate shows a linear dependence onto tunneling current, indicating that the interchange is induced via a single-electron process. For comparison, the rate at $V_s = 24$ mV is also shown in Fig. 5.8d, which shows no dependence on tunneling current. The same dependence is observed at the negative bias region. The above results indicate the enhancement of the

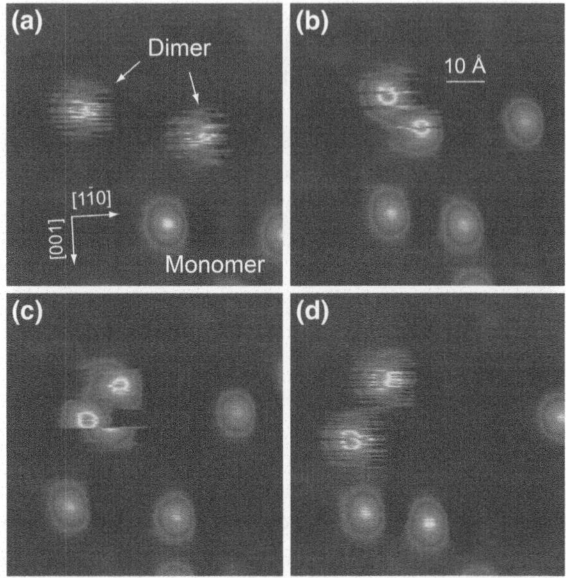

Fig. 5.9 Sequential images of two $(H_2O)_2$ (71 × 71 Å^2). **a** Two dimers show the bi-stable fluctuation corresponding to the donor–acceptor interchange. **b** As they migrate and come to close each other, one of the two configurations is strongly preferred, indicating that the interchange property is affected by the intermolecular interaction. **c** After they pass each other the preference is inverted. **d** The dimers recover their inherent dynamical properties as they diffuse out of the range of the interaction. All images were acquired at $I_t = 0.5$ nA and $V_s = 24$ mV

interchange results from the vibrational excitation of a dimer. The isotope ratio of ~1.1 on the threshold voltages suggests that the vibrational mode mainly involves hindered translation of water molecules. The DFT calculations suggested the motion of the donor-substrate stretch and the acceptor rotation is associated with the excited mode (Fig. 5.8e). By the normal-mode analysis, the corresponding vibrational energies were determined to be 36 and 34 meV for an $(H_2O)_2$ and $(D_2O)_2$, respectively. The other intermolecular modes were much lower energy. This mode directly correlates with the interchange reaction coordinate. The barrier of the interchange (0.24 eV), however, is much larger than the energy of the excited vibration, thus transferred from a tunneling electron (45 mV). This result can be explained by vibrationally-assisted tunneling process. At the excited state, the effective barrier of the interchange becomes relatively low and thin compared to that in the ground state, and thus the interchange can proceed more efficiently (indicated by red arrow in Fig. 5.8f). A significant isotope effect is also found in the interchange induced by tunneling electrons. Above the threshold, the slope of the increasing rate for an H_2O dimer is ~20 times higher than that of a $(D_2O)_2$ (note that the vertical scale in Fig. 5.8b is logarithmic), indicating vibrationally-assisted tunneling is more effective for an $(H_2O)_2$. For a gas-phase dimer, the tunneling splitting was also drastically affected by the vibrational excitation [10].

Quantum tunneling is significantly affected by surrounding conditions. Figure 5.9a–d shows a sequence of the STM images of two $(H_2O)_2$. They diffuse on the surface with the interchange. However, when they come close to each other, the fluctuation is temporarily suppressed and one of the orientations is strongly preferred (Fig. 5.9b). After they pass by each other, the preference is reversed and another configuration becomes dominant (Fig. 5.9c). Subsequently as they diffuse away from each other, their normal interchange

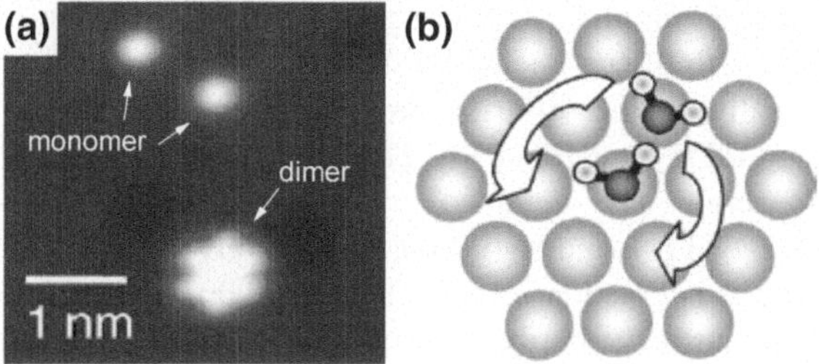

Fig. 5.10 **a** STM image of water monomer and dimer on Pt(111). **b** Schematic illustration of adsorption geometry. The six-fold symmetric "flower-like" protrusion of the dimer attributed to the free rotation of the H-bond acceptor molecule. Reprinted with permission from Ref. [16]. Copyright 2008, Elsevier

is resumed (Fig. 5.9d). The distance of ~13 Å between the interacting dimers suggests that the long-range dipole interaction affects the landscape of the interchange potential. This result highlights the sensitivity of tunneling dynamics to environmental perturbations.

5.2.4 The Impact of Substrate

The geometry of the underlying substrate has a significant impact on the dynamics of a water dimer on metal surfaces, which could provide us with more detailed knowledge. However the literature about an isolated water dimer on metal surfaces is still quite limited. Motobayashi et al. observed an isolated water dimer on a Pt(111) surface [16], where the dimer was observed as a sixfold symmetric "rosette-like" protrusion (Fig. 5.10a). This characteristic appearance is associated with the free rotation of the H-bond acceptor molecule in a dimer (Fig. 5.10b). This rotation takes place even at 6 K and extremely low bias conditions presumably due to the small barrier of rotation (calculated to be 20 meV [12]) and the rate exceeds the time-resolution of STM, resulting in the time-averaged appearance. The rosette-like dimer was also observed on Ru(0001) [17]. Unlike on Pt(111), the rotation of acceptor molecule was induced only at low gap resistance conditions ($V_s = 50$ mV, $I_t = 400$ pA) where the strong interaction between the STM tip and molecule is expected. On the other hand, in both on Pt(111) and Ru(0001) the thermal diffusion is restricted while the dimer diffuses on a Cu(110) surface even at 6 K and extremely low bias conditions ($V_s < 30$ mV). The diffusion of a water dimer on Pt(111) or Ru(0001) can be induced by the vibrational excitation via IET processes.

What results in the above dynamical differences? First the difference between Pt(111) or Ru(0001) and Cu(110) surfaces is the symmetry. Concerning the rotation of the acceptor molecule, it is much more feasible on Pt(111) or Ru(0001) than Cu(110) because the higher symmetry makes it easy to find equivalent sites having the potential minimum, resulting in the lower barrier of the rotation. Secondly, the strength of the interaction between water and metal surface affect dynamics, especially on the diffusion process, which is expected as following: Au < Ag < Cu < Pd < Pt < Ru < Rh. For instance, based on DFT calculations the adsorption energies are predicted to be 0.24, 0.33, 0.35, and 0.38 eV for Cu(111), Pd(111), Pt(111), and Ru(0001), respectively [18]. It is deduced that the strong water-surface interaction on Pt(111) or Ru(0001) pins the donor molecule and suppresses the intrinsic motion of water dimer, while the mediate interaction on Cu(110) brings it into prominence.

5.3 Summary

The donor–acceptor interchange within a water dimer isolated on a Cu(110) surface was imaged and characterized using STM. The interchange can be observed as two states fluctuating image. Using the time-resolved measurement the voltage and current dependence of the interchange rate were investigated. It was found that there is no voltage and current dependence below $V_s = 40$ mV, indicating the interchange is not induced by STM, thus intrinsic to a water dimer. Furthermore the large isotope effect ($\sim$60) was quantified in the intrinsic interchange rate between $(H_2O)_2$ and $(D_2O)_2$, suggesting that quantum tunneling is involved in the process. The excitation of the donor-substrate stretch mode was found to effectively assist the interchange tunneling, highlighting the sensitivity of the interchange dynamics to the displacement of oxygen atoms. The structure and the interchange pathway were determined by DFT calculations. The barrier calculated for the interchange (0.24 eV) cannot be overcome by mere thermal activation at 6 K, supporting the experimental results that the interchange takes place via tunneling.

References

1. T.R. Dyke, J. Chem. Phys. **66**, 492 (1977)
2. T.R. Dyke, K.M. Mack, J.S. Muenter, J. Chem. Phys. **66**, 498 (1977)
3. J.A. Odutola, T.R. Dyke, J. Chem. Phys. **72**, 5062 (1980)
4. J.A. Odutola, T.A. Hu, D. Prinslow, S.E. O'dell, T.R. Dyke, J. Chem. Phys. **88**, 5352 (1988)
5. R.S. Fellers, C. Leforestier, L.B. Braly, M.G. Brown, R.J. Saykally, Science **284**, 945 (1999)
6. K.L. Busarow, R.C. Cohen, Geoffrey A. Blake, K.B. Laughlin, Y.T. Lee, and R.J. Saykally. J. Chem. Phys. **90**, 3937 (1989)
7. G.T. Fraser, Int. Rev. Phys. Chem. **10**, 189 (1991)
8. E. Zwart, J.J. ter Meulen, W. Leo Meerts, L.H. Coudert, J. Mol. Spectrosc. **147**, 27 (1991)
9. B.J. Smith, D.J. Swanton, J.A. Pople, H.F. Schaefer, L. Radom, J. Chem. Phys. **92**, 1240 (1990)
10. N. Pugliano, J.D. Cruzan, J.G. Loeser, R.J. Saykally, J. Chem. Phys. **98**, 6600 (1993)
11. T. Mitsui, M.K. Rose, E. Fomin, D.F. Ogletree, M. Salmeron, Science **297**, 1850 (2002)

12. V.A. Ranea, A. Michaelides, R. Ramírez, P.L. de Andres, J.A. Vergés, D.A. King, Phys. Rev. Lett. **92**, 136104 (2004)
13. S. Meng, E.G. Wang, S. Gao, Phys. Rev. B **69**, 195404 (2004)
14. A. Michaelides, K. Morgenstern, Nature Mat. **6**, 597 (2007)
15. R.L. Redington, J. Chem. Phys. **113**, 2319 (2000)
16. K. Motobayashi, C. Matsumoto, Y. Kim, M. Kawai, Surf. Sci. **602**, 3136 (2008)
17. A. Mugarza, T. Shimizu, D.F. Ogletree, M. Salmeron, Surf. Sci. **603**, 2030 (2009)
18. L. Arnadottir, E.M. Stuve, H. Jonsson, Surf. Sci. **604**, 1978 (2010)

Chapter 6
Water Clusters: Formation of One-Dimensional Water Clusters

Abstract I describe the assembly and characterization of small water clusters on a Cu(110) surface in this chapter. One-dimensional (1-D) chain-like clusters (3–6 molecules), where water molecules are arranged along the Cu row to form a "ferroelectric" zigzag configuration, are assembled in a fully controlled fashion using the STM manipulation. H-bond rearrangements in the clusters are induced by a voltage pulse of STM. In addition to the chain configuration, several structural isomers are observed in larger clusters than tetramer. DFT calculations suggest a chain configuration is preferred for trimer in spite of the reduced number of H bond, while a cyclic configuration in which the number of H-bond is maximized becomes more favorable for tetramer, indicating the crossover of the stability from a chain to cyclic configuration.

Keywords Small water clusters · H-bond rearrangement · STM manipulation

6.1 Introduction

A water cluster is a discrete H-bonded assembly of water, which is of particular interest to understanding anomalous water characteristics. Water clusters are considered as a model system to examine the many-body, or cooperative nature of H-bonding interaction. Terahertz laser vibration–rotation-tunneling and mid-IR spectroscopy have been employed to observe small water clusters (2–6 molecules) [1]. Numerous semi-empirical and ab initio quantum mechanical calculations on water clusters have also been reported [2]. For trimer to pentamer, it was found to form a cyclic structure where each monomer acts both as a single donor and acceptor of H-bond.

The structure and growth mechanism of water clusters on metal surfaces are related to various chemical processes such as wetting, corrosion, and heterogeneous

T. Kumagai, *Visualization of Hydrogen-Bond Dynamics*, Springer Theses,
DOI: 10.1007/978-4-431-54156-1_6,

catalysis [3–6]. Although structural characterizations of two-dimensional (2-D) water layers on metal surfaces have been extensively studied in the last decades, there still remain some open questions in small water clusters. This is because a suitable experimental technique to characterize such clusters has been lacking. As described in Chap. 1, vibration spectroscopy has been employed to observe water clusters and islands. For instance, by careful analysis of infrared spectra, Yamamoto et al. showed monomeric adsorption and aggregation into small clusters, 2-D islands and finally a bulk amorphous ice layer on a Rh(111) surface at 20 K [7]. However the assignment of vibrational spectra for water clusters adsorbed on metal surfaces is not straightforward and there remains a considerable ambiguity in band assignment to different structures.

Recently, a low-temperature STM (LT-STM) have been used to address the direct imaging of single water molecules and its aggregation process on metal surfaces. Salmeron and co-workers investigated the adsorption and aggregation of water on Pd(111) using LT-STM [8–10]. They directly observed the diffusion of water monomer and small clusters across the surface by acquiring a sequence of images at 40 K [8]. Interestingly, it was found that the diffusion rate dramatically depends on the cluster size and determined to be 2.3×10^{-3}, 50 and 1.02 Ås^{-1} for the monomer, dimer and trimer/tetramer, respectively. The anomaly of the dimer was theoretically explained by Ranea et al. They attributed it to the existence of tunneling path in the diffusion process of the dimer (11; detailed in Chap. 5). In general, larger clusters are expected to have a lower mobility as they can bind to the surface commensurately via the oxygen atoms of alternate water molecules in the H-bonded structures. Increasing water coverage on a Pd(111) surface at 40 K led to the formation of clusters of various shapes/sizes and finally the hexagonal rings growth into ordered honeycomb structures with a local periodicity. In the series of papers, Morgenstern and co-workers investigated the nucleation, growth and structure of small water clusters adsorbed on Cu, Ag and Au (111) surfaces. [12–19].These surfaces are inert and no extended 2-D structures are found on the Cu and Ag(111) surfaces. Continuous improvements in experimental procedures and techniques of STM have led to high resolution STM images and an improved understanding of cluster formation. The cyclic configuration of a water hexamer was imaged on Cu(111) and Ag(111) surfaces [18], which is a special case in cluster chemistry as its cyclic form represents the basic structural unit that builds ice. The cyclic hexamer was sequentially decorated by additional water molecules and yielded a heptamer, octamer, and nonamer on the surface.

Along with LT-STM experiments, theoretical efforts have been devoted to predict the structure of small clusters. As shown in Figs. 6.1 and 6.2, a cyclic configuration was proposed as the most stable structure on close-packed metal surfaces due to the maximized number of H bond [20, 21]. The calculations predicted the cyclic water hexamer is the lowest energy structure consisted of six water molecules where each molecule acts as a single proton donor/acceptor. The maximization of the H-bond number is a fundamental tendency for gas phase clusters. However, the presence of surface sometimes causes a unique growth of water. For instance, 1-D chain ice was observed on the terrace of Cu(110) [22, 23];

Fig. 6.1 Water monomer and small clusters adsorbed on a Pt(111) surface. Reprinted with permission from Ref. [20]. Copyright 2004, American Physical Society

when water molecules are adsorbed onto Cu(110) at ~80 K, they spontaneously aggregate into the 1-D structure with its width of 10 Å, which is aligned along [001] axis. Theoretical calculations predicted that the 1-D structure consists of interlinked water pentagon units [23]. The water growth along the Pt step edges also shows a zigzag H-bond chain [24]. This class of 1-D structure is of particular interest, because such a 1-D configuration is known to exist in biological systems, mediating proton translocation across the transmembrane proteins [3]. Water chains confined in carbon nanotubes were intensively studied as a model system to investigate the structure and dynamical properties [4, 5].

As described above, several types of water clusters and the growth of extended structures have been revealed, but the structural characterization of intermediate size clusters (3–5 molecules) still remains lacking.

6.2 Results and Discussions

6.2.1 Water Trimer on Cu(110)

A water trimer is produced by the reaction between a water monomer and dimer. Figure 6.3 shows the production of a water trimer. A monomer and dimer are characterized as a round and bi-stable fluctuating images, respectively, and they diffuse on the surface even at 6 K. A trimer is formed spontaneously when they sufficiently come close to each other. A trimer is characterized by an asymmetric triangle protrusion and never dissociated into a monomer and dimer spontaneously at the conditions of its production, indicating it is energetically favorable. It is found that four orientations exist for a trimer, which can be interconverted with a voltage pulse of STM (Fig. 6.4a–d). The threshold voltage was ~0.12 V.

The structure of trimer is optimized by DFT calculations[1] [shown in the bottom of the STM images in Fig. 6.4]. In the optimized structure water molecules aligns

[1] DFT calculations were performed using the STATE code [Y. Morikawa et al. Phys. Rev. B 69, 041403 (2004).], within the Perdew–Burke–Ernzerhof (PBE) [J. P. Perdew et al. Phys. Rev. Lett. 77, 3865 (1996).] generalized gradient approximation (GGA). Wave functions and argumentation charge were expanded in a plane-wave basis set, and electron–ion interactions were described by

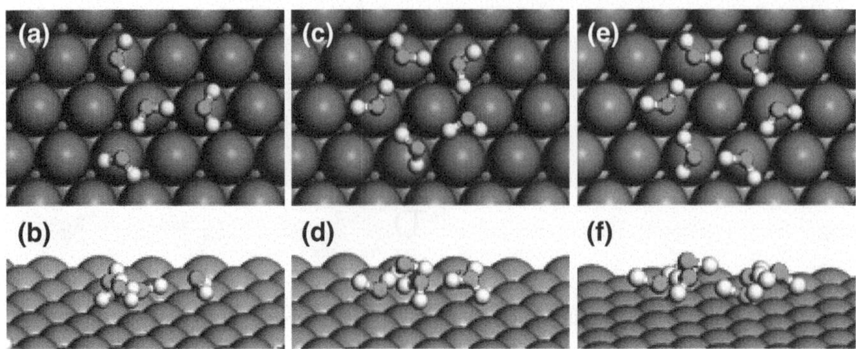

Fig. 6.2 Water tetramer (**a**)–(**b**), pentamer (**c**)–(**d**), and hexamer (**e**)–(**f**) on a Cu(111) surface. Reprinted with permission from Ref. [21]. Copyright 2007, RSC Publishing

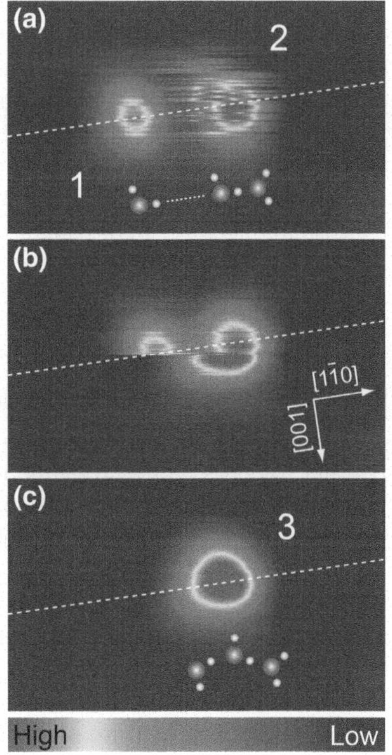

Fig. 6.3 Sequential STM images of the formation of a water trimer. **a** A monomer and dimer are imaged. **b** They diffuse on a Cu(110) surface at 6 K and associate spontaneously when the two species sufficiently come close to each other. **c** The produced trimer appears as a triangle-shaped protrusion. The STM images were acquired at $V_s = 24$ mV and $I_t = 0.5$ nA. The image size is 30 × 50 Å. The white dashed line in the images represents the atomic row of Cu(110)

(Footnote 1 continued)
pseudopotentials. In the calculations the trimers were put on one side of five-layer slabs with (3 × 4) periodicities and artificial electrostatic interaction was eliminated by the effective screening medium method [M. Otani and O. Sugino, Phys. Rev. B 73, 115407 (2006); I. Hamada et al. ibid. 80, 165411 (2009).]. Brillouin zone sampling was done using 4 × 4 Monkhorst–Pack [H. J. Monkhorst and J. D. Pack, Phys. Rev B 13, 5188 (1976).] k-point sets for the cell.

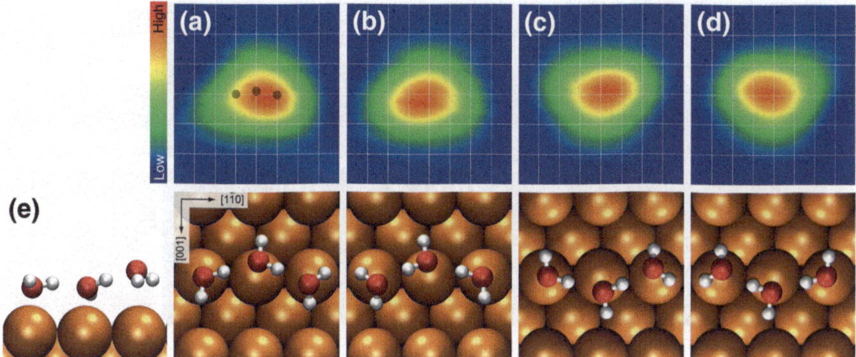

Fig. 6.4 High-resolution STM images (top) and calculated structure (bottom) of an H_2O trimer. White grid lines in the images represent the lattice of Cu(110). A trimer can be converted among the four orientations. (a) A trimer has an asymmetric character in the image. The chain configuration is obtained as the optimal structure and water molecules are linked via H bonds. Black filled circles in the STM image indicate the positions of a water molecule. (b)-(d) Other orientations. All images were acquired at $V_s = 24$ mV and $I_t = 0.5$ nA. The size is 23×23 Å^2

along the Cu atomic row direction (labeled as [1$\bar{1}$0]-trimer) and the calculated binding energy is 0.458 eV/H_2O with respect to monomers isolated in the gas phase. The H-bond acceptor molecule is higher than the donor by 0.76 Å. Therefore, the acceptor is expected to be imaged as a higher protrusion in STM, giving rise to an asymmetric character and the conversion of the image corresponds to the rearrangement of H bond within a trimer.

Another type of trimer is produced by the association between a monomer and a dimer which are located on the adjacent Cu atomic rows. Figure 6.5a shows two different types of trimer. No other structural isomer was observed for trimer on Cu(110). The second type of trimer also shows four orientations (Fig. 6.5b–e). The orientation is reversibly flipped between (b) and (c) or (d) and (e) spontaneously even at low bias conditions ($V_s < 30$ mV) with the time scale of ~ 1 s. The second type of trimer eventually transformed into a [1$\bar{1}$0] trimer, suggesting that it is a metastable configuration. The plausible structure is proposed in Fig. 6.6f. In the optimized structure the two water molecules are bonded via H bond and across the trough along [001] axis, and another is attached to the end as the H-bond acceptor (label this as [001]-trimer). Although a cyclic structure is the most stable in a gas phase trimer as well as on close-packed surfaces, it is not observed on Cu(110). Probably the cyclic configuration cannot find an optimum adsorption site on Cu(110) because of the geometrical constraint of the quasi-1-D structure. For the same reason the one-donor and two-acceptor configuration proposed as the stable configuration on Cu(111) [18, 21] and Ni(111) [25] is unfavorable on Cu(110). These findings suggest a decisive role of the underlying substrate in the clustering behavior of water molecules. For both of trimers, diffusion is never observed at 6 K.

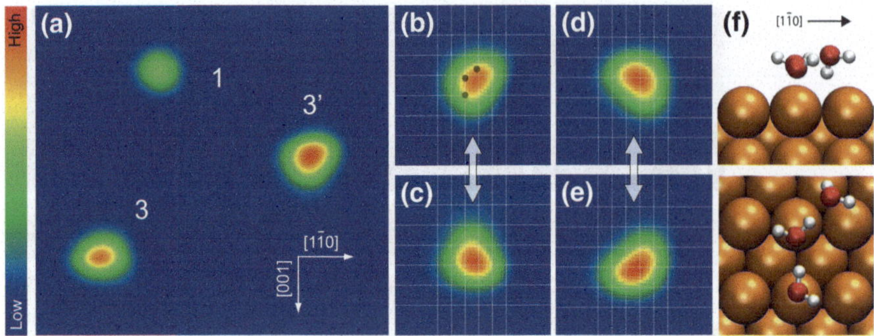

Fig. 6.5 **a** In addition to a [1$\bar{1}$0]-chain (*left bottom*, labeled with 3), another type of trimer (*right middle*, labeled with 3′) is observed. The latter trimer is flipped spontaneously between **b** and **c** or **d** and **d** with the time scale of ~1 s even at low bias voltages, and eventually converts into a [1$\bar{1}$0]-chain. The apparent height is 0.78 Å for the second trimer. **f** The optimized structure for a [001]-trimer, where the two water molecules are bonded along [001] direction and another is attached to the end, forming a chain-like configuration. The images were obtained at $V_s = 24$ mV and $I_t = 2$ nA for (**a**), $V_s = 24$ mV and $I_t = 0.5$ nA for (**b**)–(**e**). Scan area is 75 × 68 Å^2 for (**a**), 28 × 28 Å^2 for (**b**)–(**e**)

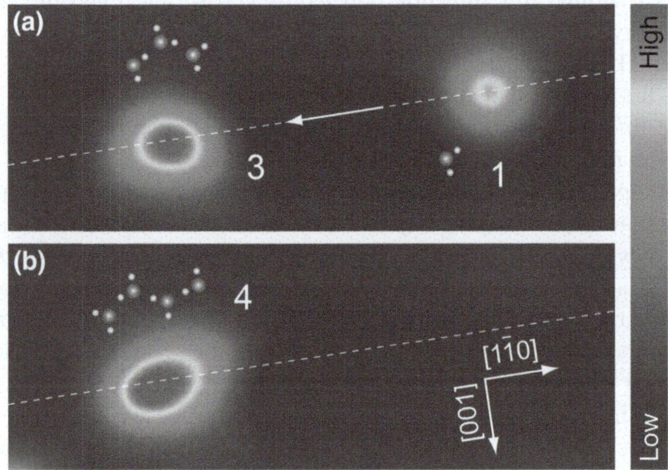

Fig. 6.6 Sequential STM images of the formation of a chain-like tetramer. **a** The monomer is manipulated and attached to the end of the [1$\bar{1}$0]-chain trimer. **b** The produced tetramer is imaged as a distorted parallelogram. The schematic models are illustrated in the images. The direction of H bond was changed during the reaction. The images were obtained at $V_s = 24$ mV and $I_t = 0.5$ nA. Scan area is 69 × 25 Å^2

6.2.2 *Water Tetramer on Cu(110)*

A [1$\bar{1}$0]-trimer can be lengthened using the STM manipulation (Fig. 6.6). When a monomer is attached to the end of a [1$\bar{1}$0]-trimer, a parallelogram protrusion is

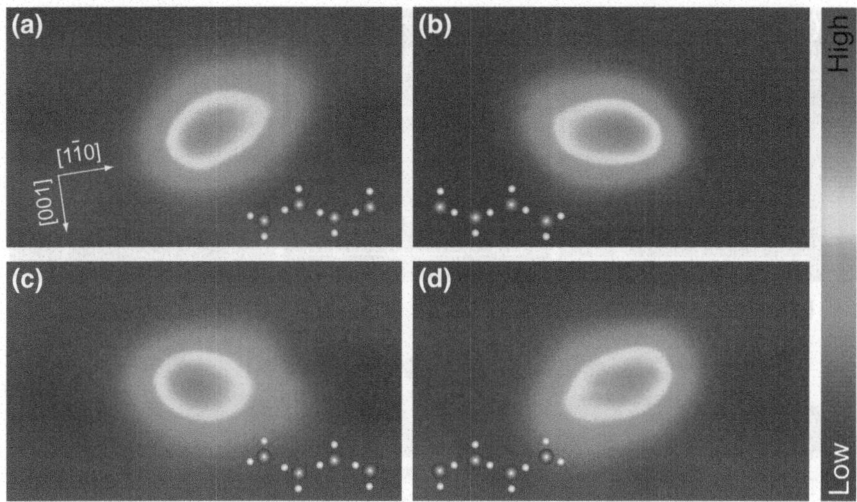

Fig. 6.7 a–d STM images of the four orientations of a chain-like tetramer. The conversion is induced with a voltage pulse of the STM. The schematic models are illustrated in the images. The images were obtained at $V_s = 23$ mV and $I_t = 0.5$ nA. Scan area is 37×22 Å^2

formed. A distortion in the image appearance, which is also observed for a [1$\bar{1}$0]-trimer, is discernible and there are four orientations as shown in Fig. 6.7a–d. In addition to the chain configuration, several structural isomers emerge due to the increased degrees of freedom for a tetramer. The conversion into other isomers can be induced with a voltage pulse of STM. The voltage pulse over a chain-like tetramer induces the conversion of a parallelogram into a triangle- or round-shaped tetramer (Fig. 6.8b, c). These three structural isomers can be mutually converted. The threshold voltages of the conversions are estimated to be ~120, ~140 and ~180 mV for parallelogram, triangle, and round-shaped tetramers, respectively, with the tip fixed at $V_s = 23$ mV and $I_t = 0.5$ nA over a tetramer.

Several optimized structures of tetramer were determined by DFT calculations.[2] The optimized structures are shown below the STM images in Fig. 6.8. Similar to trimer, the H-bond acceptor molecule located at the end is higher than the others in the chain configuration, which contributes to the brightest protrusion in the image. The vertical displacement between the H-bond acceptor and donor located at the opposite end is 0.74 Å. Two more stable structures are determined; one has a cyclic configuration in which each molecule is linked via H bond. A round-shaped tetramer is assigned to the cyclic configuration. The other forms a "tetrahedral" configuration in which the center molecule simultaneously acts as a double donor and a single acceptor of H bond. The similar configuration was proposed for a tetramer on Cu(111) [12]. A triangle-shaped tetramer is assigned to this

[2] The water tetramers were put on one side of five-layer slabs with 4×6 periodicities and Brillouin zone sampling was done using 2×2 Monkhorst–Pack k-point sets for the cell.

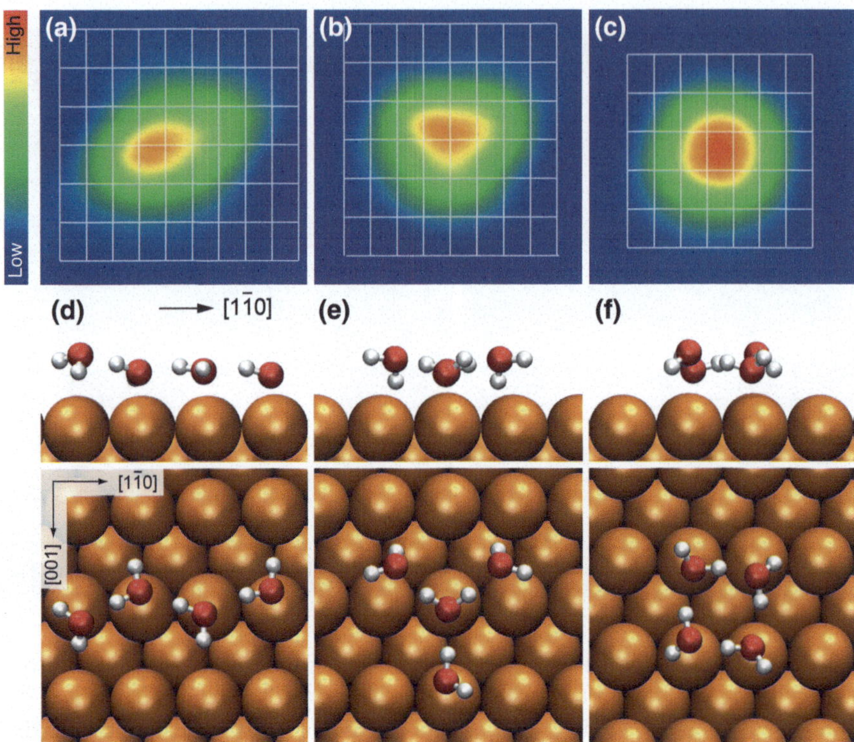

Fig. 6.8 STM images of three isomers for a tetramer and the proposed structures obtained by DFT calculations. **a** The optimized structure (**d**) is analogous to that of a chain-like trimer. **b** The triangular species is assigned to the tetrahedral configuration (**e**) with the center molecule acting as a double donor and single acceptor. **c** The round species is centered on the hollow site, and is assigned to the cyclic configuration (**f**). The apparent height (maximum) is 0.8, 0.8 and 1.0 Å for the chain (**a**), triangular (**b**) and round (**c**) tetramers, respectively. All STM images were obtained at V_s = 24 mV and I_t = 2 nA. The image size is 25 × 25 Å^2

configuration. It is noted that individual molecules in the cyclic configuration has different height from the surface, but the STM image is fully symmetric. Whereas the height difference of water molecules is clearly reflected in the appearance for a chain-like trimer and tetramer, it is not observed for a cyclic tetramer. The absence of the expected corrugation was also reported for a cyclic hexamer on Cu(111) [18]. It is attributed either to a tip-induced reorientation of water molecules within the cluster as they are observed, or alternatively to a dynamical fluctuation among several configurations that is much faster than the time resolution of the STM.

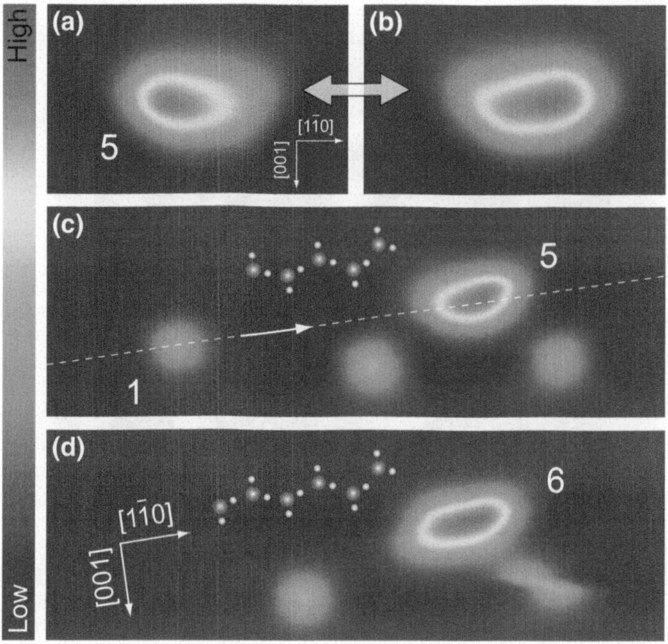

Fig. 6.9 **a** STM image of a chain-like pentamer. **b** The counterpart of (**a**) in which the sequential H bonds point in the opposite direction. The acceptor molecule at the end in the chain appears brightest, and thus we can distinguish between the two directions. **c** A water monomer and pentamer located on the same atomic row (*dashed line*). **d** The monomer is attached to one end of the pentamer to yield a chain hexamer. The corresponding chain structures are depicted in the inset. All images were obtained at $V_s = 24$ mV and $I_t = 0.5$ nA. The image size is 39×19 Å^2 for (**a**) and (**b**), 99×33 Å^2 for (**c**) and (**d**)

6.2.3 Larger Clusters on Cu(110)

A chain-like pentamer can be produced with the same manner as the tetramer. A chain-like pentamer is characterized by a trapezoidal protrusion (Fig. 6.9a, b). A distorted character is still discernible in the STM appearance and it can be also converted among the other orientations. Moreover, in order to produce a chain-like hexamer, a monomer located on the same Cu atomic row was attached to the end of a chain-like pentamer [Fig. 6.9c, d). The hexamer is imaged as a parallelogram protrusion. The dependence of the STM appearances on the odd/even constituent numbers indicates the zigzag arrangement of water molecules along the chains. Furthermore the orientation of H bonds in the chains is expected to be the illustration shown in Figs. 6.9c, d, suggesting that the chains are "ferroelectric" where one OH bond of each molecule contributing to the H bonds in one direction. Analogous 1-D water chains were proposed to grow along the step edge of Pt surface [20, 24, 26, 27]. This type of zigzag arrangement of water molecules was verified by the x-ray diffraction [27], whereas the H-bond configuration remained

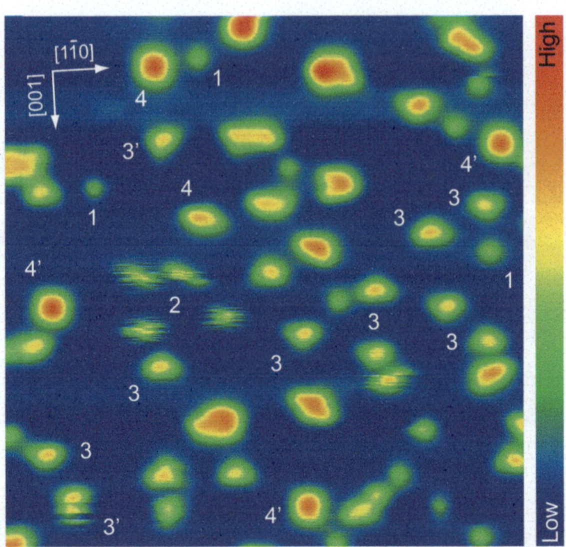

Fig. 6.10 STM images of various clusters spontaneously formed at ~30 K. The image was acquired at 6 K. The identified clusters are labeled with their constituent numbers. The majority species of trimer is a [1$\bar{1}$0]-chain. The meta-stable trimer (labeled with 3′) in the *bottom left* was mobile and appeared fractional. Tetramer is most likely observed in the cyclic form (labeled with 4′). The range of the height shown in the color bar is from −0.2 to 1.3 Å. The image was obtained at $V_s = 24$ mV and $I_t = 0.5$ nA, and the scan size is 170 × 170 Å^2

open question because H atoms were invisible with x-ray. A chain configuration is not converted into the other configurations spontaneously at low bias conditions, suggesting that it is sufficiently stable. It would appear that an anisotropic atom arrangement of Cu(110) allows water molecules to arrange one-dimensionally despite the reduction of the number of H bond.

The temperature dependence of clustering behavior provides us with insights of the thermodynamics. As the temperature increases, water molecules readily diffuse and aggregate spontaneously to form a variety of clusters. Figure 6.10 is a typical STM image after the surface is heated up to ~30 K. Most of the clusters can be identified and the cluster size is labeled in the image. A trimer and tetramer are observed mainly as a [1$\bar{1}$0]-chain and cyclic form, respectively, suggesting that they are thermodynamically favorable among several isomeric structures. Also longer chains as well as unidentified larger clusters are observed. It is clear that the trimer is the most predominant species, indicating that the diffusion of the trimer is still restricted at 30 K.

6.3 Summary

Small water clusters were assembled and imaged on Cu(110) using STM. The structure was determined with the help of DFT calculations. Water molecules can be arranged one-dimensionally along the atomic row of Cu(110), forming stable H-bonded chain clusters in which water molecules are arranged along the Cu row to form a "ferroelectric" zigzag configuration. The direction of H-bond in the chain was rearranged by a voltage pulse of STM. Despite the reduction of the number of H bond, the trimer prefers chain-like configurations to a cyclic form due to the geometrical constraint of the quasi-1-D structure of Cu(110). For the tetramer, however, a cyclic form was found to be the most stable configuration among the structural isomers including the chain-like one. These inclinations are quite different from those of gas-phase counterparts and on other surfaces, demonstrating the impact of the substrate on the clustering behavior of adsorbed water molecules.

References

1. F.N. Keutsch, R.J. Saykally, Proc. Natl. Acad. Sci. USA **98**, 10533 (2001)
2. R. Ludwig, Angew. Chem. Int. Ed. **40**, 1808 (2001)
3. M.H.B. Stowell, T.M. McPhillips, D.C. Rees, S.M. Soltis, E. Abresch, G. Feher, Science **276**, 812 (1997)
4. G. Hummer, J.C. Rasaiah, J.P. Noworyta, Nature **414**, 188 (2001)
5. D.J. Mann, M.D. Halls, Phys. Rev. Lett. **90**, 195503 (2003)
6. A. Verdaguer, G.M. Sacha, H. Bluhm, M. Salmeron, Chem. Rev. **106**, 1478 (2006)
7. S. Yamamoto, A. Beniya, K. Mukai, Y. Yamashita, J. Yoshinobu, J. Phys. Chem. B **109**, 5816 (2005)
8. T. Mitsui, M.K. Rose, E. Fomin, D.F. Ogletree, M. Salmeron, Science **297**, 1850 (2002)
9. J. Cerda, A. Michaelides, M.L. Bocquet, P.J. Feibelman, T. Mitsui, M. Rose, E. Fomin, M. Salmeron, Phys. Rev. Lett. **93**, 116101 (2004)
10. J.M. Blanco, C. Gonzalez, P. Jelinek, J. Ortega, F. Flores, R. Perez, M. Rose, M. Salmeron, J. Mendez, J. Wintterlin, G. Ertl, Phys. Rev. B **71**, 113402 (2005)
11. V.A. Ranea, A. Michaelides, R. Ramirez, P.L. de Andres, J.A. Verges, D.A. King, Phys. Rev. Lett. **92**, 136104 (2004)
12. K. Morgenstern, Surf. Sci. **504**, 293 (2002)
13. K. Morgenstern, J. Nieminen, Phys. Rev. Lett. **88**, 066102 (2002)
14. K. Morgenstern, J. Nieminen, J. Chem. Phys. **120**, 10786 (2004)
15. K. Morgenstern, K.H. Rieder, J. Chem. Phys. **116**, 5746 (2002)
16. H. Gawronski, K. Morgenstern, K.-H. Rieder, Eur. Phys. J. D **35**, 349 (2005)
17. K. Morgenstern, H. Gawronski, T. Mehlhorn, K.H. Rieder, J. Mod. Opt. **51**, 2813 (2004)
18. A. Michaelides, K. Morgenstern, Nat. Mat. **6**, 597 (2007)
19. M. Mehlhorn, J. Carrasco, A. Michaelides, K. Morgenstern, Phys. Rev. Lett. **103**, 026101 (2009)
20. S. Meng, E.G. Wang, S. Gao, Phys. Rev. B **69**, 195404 (2004)
21. A. Michaelides, Faraday Discuss. **136**, 287 (2007)
22. T. Yamada, S. Tamamori, H. Okuyama, T. Aruga, Phys. Rev. Lett. **96**, 036105 (2006)
23. J. Carrasco, A. Michaelides, M. Forster, S. Haq, R. Raval, A. Hodgson, Nat. Mat. **8**, 427 (2009)

24. M. Morgenstern, T. Michely, G. Comsa, Phys. Rev. Lett. **77**, 703 (1996)
25. D. Sebastiani, L. Delle Site, J. Chem. Theory Comput. **1**, 78 (2005)
26. M.L. Grecea, E.H.G. Backus, B. Riedmüller, A. Eichler, A.W. Kleyn, M. Bonn, J. Phys. Chem. B **108**, 12575 (2004)
27. M. Nakamura, N. Sato, N. Hoshi, J.M. Soon, O. Sakata, J. Phys. Chem. C **113**, 4538 (2009)

Chapter 7
Hydroxyl Group: Tunneling Dynamics of Hydrogen Atom

Abstract I describe the production and characterization of isolated hydroxyl species on a Cu(110) surface in this chapter. A hydroxyl group can be produced by the STM-induced dissociation of a water molecule. A hydroxyl can be further dissociated into atomic oxygen. It is found that a hydroxyl has an inclined geometry against the surface normal and switches back and forth between the two orientations via the H atom tunneling. The switching results in the characteristic paired depression aligned along the [001] direction in the STM appearance of hydroxyl. The tunneling switching can be observed directly for a deuterated species (OD) within the time-resolution of STM, while it is smeared out for an OH due to a significant increase of the tunneling rate. The switching is enhanced by the vibrational excitation of the OH(OD) bending mode which is directly associated with the switching reaction coordinate.

Keywords Tunneling dynamics of hydrogen atom · Hydroxyl groups on metal surfaces · Single molecule dissociations · STM-IETS

7.1 Introduction

Quantum effects of H atom are of fundamental importance in diverse chemical, physical, and biological processes [1–5]. Especially, tunneling in a two-state system is a key interest because it is considered as one of the most ubiquitous questions in physics and chemistry. For instance, the flipping of an isolated NH_3 molecule is the simplest model of H-atom tunneling, which was investigated by Hund in the early days of quantum mechanics [6].

In surface science, quantum dynamics of H/proton has been investigated in connection with the elementary processes of heterogeneous catalysis, energy production and storage, namely fuel cells. The light mass of H atom results in the unique behavior even at metal surfaces. For instance, Christmann et al. proposed

T. Kumagai, *Visualization of Hydrogen-Bond Dynamics*, Springer Theses,
DOI: 10.1007/978-4-431-54156-1_7,

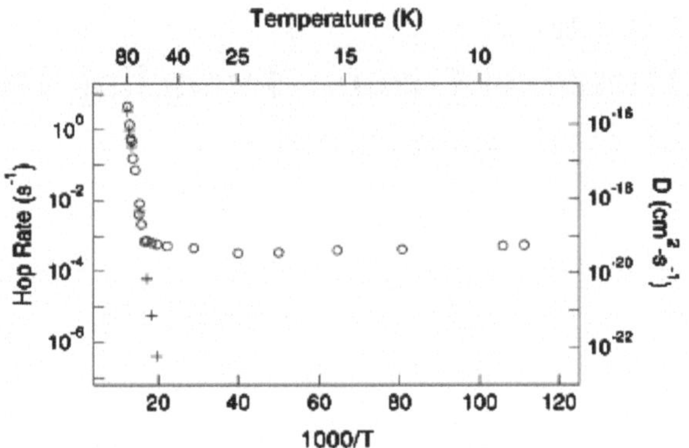

Fig. 7.1 Diffusion rate of H and D on Cu(001) as a function of temperature (Arrhenius plot) between 80 and 9 K measured by atom tracking technique. The hopping rates of H and D are represented by *open circle* and cross, respectively. Reprinted with permission from Ref. [13, 14]. Copyright 2000, American Physical Society

the quantum delocalization of H [7] and “the H-band model”, where H is described in terms of the two-dimensional atomic energy bands. The first theoretical study on the quantum motion of chemisorbed hydrogen was reported by Puska et al. in 1983 [8]. The H-band model was experimentally examined by Mate and Somorjai in 1986 [9], in which they showed the evidence of the quantum delocalization in the study of H atom on Rh(111) using high resolution electron energy loss spectroscopy (HREELS). Quantum-delocalization of H atom on transition-metal surfaces was summarized by Nishijima et al. [10].

Diffusion is a common process of H atom on metal surfaces. The diffusion behavior was first investigated by means of field emission microscopy in 1957 [11, 12]. More recently, the unique dynamics of H was directly observed on a metal surface using variable temperature STM. In 2000 Lauhon and Ho showed the clear evidence of quantum diffusion of H atom on Cu(100) [13, 14]. They investigated the temperature dependence of the diffusion rate using the atom tracking technique of STM. They showed an explicit deviation from Arrnenius equation and the rate becomes plateau below 60 K (Fig. 7.1). The diffusion was described as incoherent tunneling in the presence of lattice and electronic excitations (electron and phonon scattering cause decoherence of the H-atom wave function and particle localization). Interestingly, it was found that tunneling plays a crucial role even in heavy atoms and molecule [15–17], where the tunneling of Cu and Co atoms and CO molecule was characterized on metal surfaces by low-temperature STM. These results suggested that the tunneling is feasible even for heavy particles within a reduced potential in the atomic-scale distances.

?A3B2 twb=.25w?>Hydroxyls on surfaces have been discussed in conjunction with the structure of a first wetting layer [18, 19] and H/proton transfer processes [20]. In general, a spontaneous dissociation of water molecules on transition metal surfaces is not feasible at low temperature. However, the situation changes considerably as following cases; (i) a surface is heated and thermally activated processes takes place. (ii) Pre-adsorbed oxygen atoms exist on the surface. For instance, the water dissociation was observed on the oxygen-covered Pt(111)

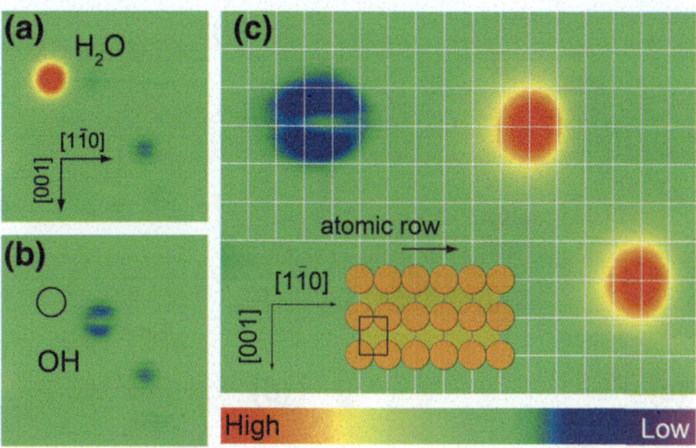

Fig. 7.2 **a** A voltage pulse of 2 V for 0.5 s was applied to a water molecule imaged as a round protrusion. The tip was fixed at the set point of $I_t = 0.5$ nA and $V_s = 24$ mV during the pulse. **b** After the pulse the product (OH) appears as paired depressions. The *black circle* represents the original position of the parent water molecule. The image size is 47 × 47 Å^2. **c** The relative position of OH species to nearby water molecules which is bonded to a top of Cu atom. The white grid lines indicate the lattice of the Cu(110) surface, indicating that the adsorption site of OH is the twofold short-bridge site. The image size is 46 × 30 Å^2. All images were acquired at $V_s = 24$ mV and $I_t = 0.5$ nA

[19, 21, 22], Ir(110) [23] and Cu(110) [24] surfaces. (iii) A surface is irradiated by an electron or UV beam [25].

7.2 Results and Discussions

7.2.1 STM-Induced Dissociation of a Water Molecule on Cu(110)

A hydroxyl group can be produced by the dissociation of a water molecule with a voltage pulse of STM (Fig. 7.2). First the STM tip is fixed over the water molecule (the round protrusion at the upper left) in Fig. 7.2a and then a voltage pulse of 2 V is applied. After the pulse, the water molecule changes into paired depression aligned along the [001] direction in Fig. 7.2b, which is assigned to a hydroxyl group. The dissociation competes with the hopping of a parent water molecule, and a hydroxyl is usually produced away from the original position. The dissociation can be induced with a probability of ~50 % by a pulse of 2 V for 0.5 s. An STM-induced water dissociation was observed on Cu(100) [26], Cu(111) [27], Ru(0001) [28] and MgO film on a silver surface [29]. The dissociation is induced via the vibrational excitation or electronic excitation of a water molecule. As described in the previous chapter, water molecules are adsorbed on a top of Cu atoms. Accordingly, the adsorption site of a hydroxyl group is determined from its relative location against the nearby water molecules to be a short-bridge site (Fig. 7.2b), which is consistent with the previous theoretical predictions [30–32].

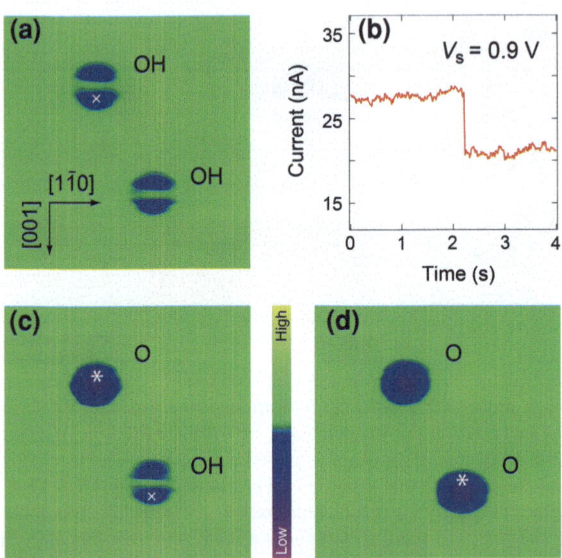

Fig. 7.3 Sequential images of the dissociation of two OH groups into atomic oxygen. **a** A voltage pulse of 0.9 V is applied over one depression (indicated by the *white cross*) of the *top-left* hydroxyl. The tip height is fixed at the set point of $I_t = 0.5$ nA and $V_s = 24$ mV. **b** The tunneling current during the voltage pulse. **c** and **d** Images taken after the pulses show the successive dissociation into round depressions which are assigned as atomic oxygen. The *asterisk* represent the short-bridge site occupied by a parent OH, which indicate that the product oxygen is displaces along [001] and located an adjacent hollow site. The image size is 42 × 42 Å^2. All images were acquired at $V_s = 24$ mV and $I_t = 0.5$ nA

The further dissociation of a hydroxyl into atomic oxygen can be induced. Figure 7.3 shows the dissociation of OH groups. The STM tip is positioned over the one depression (indicated by a white cross) in Fig. 7.3a and then a 0.9 V pulse is applied. Figure 7.3b shows tunneling current during a voltage pulse and the abrupt drop corresponds to the moment of the dissociation. After the pulse it dissociated into an atomic oxygen imaged as a round depression with the apparent height of −0.3 Å. The asterisk in Fig. 7.3b, c indicates the original position of the parent OH species (short-bridge site) and atomic oxygen is situated on the adjacent fourfold hollow site. The appearance and adsorption site of the generated oxygen are consistent with the previous result [33]. In Fig. 7.3d another OH is also dissociated. It was found that the position of the produced oxygen can be controlled depending on the initial tip position before a voltage pulse. Atomic oxygen is always formed at the hollow site under the tip. Therefore the current trace during the dissociation always shows a drop as seen in Fig. 7.3b when the dissociation takes place. The dissociation of a hydroxyl occurs at both bias polarities, and the threshold is estimated to be 0.9 (1.5) V for OH (OD) with the variation of ~0.2 V depending on the tip conditions. The hydrogen abstraction is assumed to be induced via the overtone excitation of the O–H stretch mode. The isotope effect

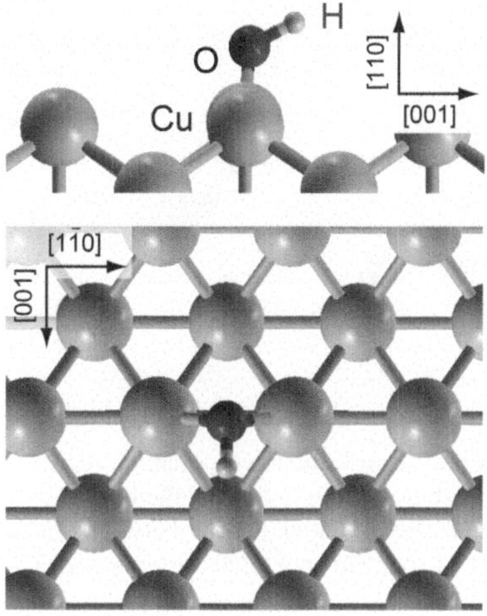

Fig. 7.4 The calculated structure of hydroxyl group on Cu(110) from side and top view. The OH bond length and the angle between the axis and the surface normal are 0.99 Å and 62°, respectively

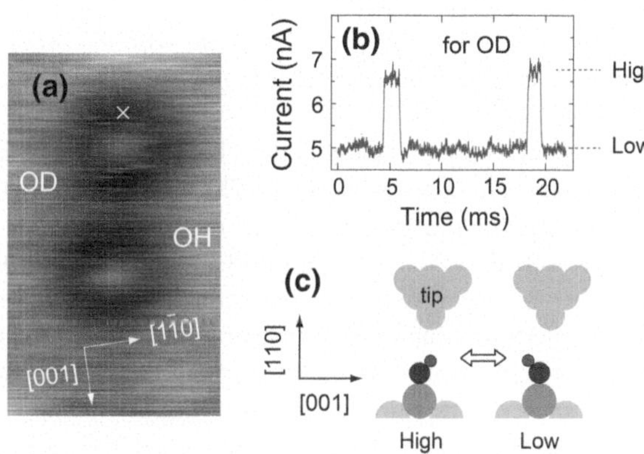

Fig. 7.5 **a** STM image simultaneously acquired for an OH and OD. **b** Tunneling current recorded with the tip fixed over the one depression OD at $I_t = 5$ nA at $V_s = 24$ mV. **c** Schematic illustration of the switching between the two orientations under the tip. The high-current (low-current) state is assigned to the orientations of OD pointing toward (away from) the tip

can be ascribed to the reduced lifetime of the vibrational excited states between OH and OD, since the lower vibrational energy of OD stretch results in a more efficient decay of the excited state into the ground state, which prevents the vibrational up-pumping process [34].

7.2.2 Structure and Dynamics of a Hydroxyl Group on Cu(110)

The structure of OH on Cu(110) was determined by DFT calculations.[1] The inclined geometry of the OH axis toward [001] axis is obtained as the optimized

[1] DFT calculations were performed using the STATE code [Y. Morikawa et al. Phys. Rev. B 69, 041403 (2004).]. The OH was put on one side of a three-layer Cu slab arrayed in a 2 × 3 surface unit cell, and the vacuum region of 12.89 Å was inserted between slabs. A GGA-optimized lattice

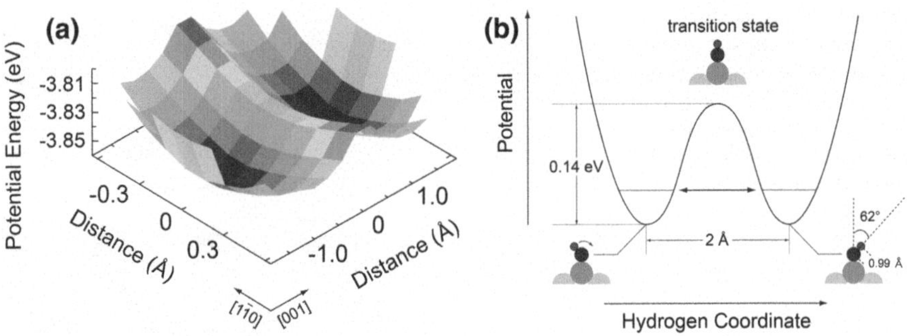

Fig. 7.6 **a** Calculated potential-energy surface of OH as a function of the lateral displacements of H atom from the top of the bridge site. The two minima correspond to the inclined geometries, and the potential barrier between the minimum and the saddle point (for switch motion) is 0.14 eV. The potential minima are 3.94 eV in depth with respect to free OH. **b** Schematic illustration of the energetically favorable pathway of the hydroxyl switching

structure (Fig. 7.4), which is consistent with the previous calculations [30, 31] and experiment [35]. The geometrical parameters are almost same as those reported previously.

There is a notable isotope effect between OH and OD. Although no remarkable difference can be seen in the STM appearance between them (Fig. 7.5a), the bi-state fluctuation can be observed in the current trace measured over OD. Figure 7.5b shows a typical current trace measured over OD while the tip is fixed over the one depression. This fluctuation was never observed for OH. Thus OH and OD are distinguishable. It is plausible the bi-stable fluctuation of OD corresponds to the switching between the two orientations as illustrated in Fig. 7.5c. The STM appearance of the paired depression aligned along [001] axis is interpreted as a time averaged image of the switching. The STM simulation based on the Tersoff-Hamann approach predicted that the side of the H atom appears brighter than the opposite position though a depression feature could not be reproduced. I tentatively assigned that the high-current (low-current) state is assigned to the orientations of OD pointing toward (away from) the tip.

To elucidate the dynamical properties of hydroxyl, the potential energy surface (PES) was calculated along the lateral coordinates of the H atom (Fig. 7.6a). In the calculations the Cu atoms in topmost two layers were optimized at each point. The PES shows a double-well structure, where the two minima correspond to the two

(Footnote 1 continued)
constant of 3.64 Å, which is 0.8 % larger than the experimental value of 3.61 Å, was used to construct the slabs. A (4 × 4) Monkhorst–Pack [H. J. Monkhorst and J. D. Pack, Phys. Rev. B 13, 5188 (1976).] *k*-point set was used to sample the surface Brillouin zone, and the Fermi level was treated by the first order Methfessel-Paxton scheme [M. Methfessel and A. T. Paxton, Phys. Rev. B 40, 3616 (1989).] with the 0.05 eV smearing width. During the structural optimization, adsorbates and two topmost Cu layers were allowed to relax until the forces on them were less than 0.05 eV/Å.

equivalent inclined geometries. At the transition state, the OH is expected to be an upright geometry. In the minimum energy path the H atom moves along the [001] axis. The estimated lowest barrier of 0.14 eV, however, is too high to be overcome via mere thermal process at 6 K, suggesting the switching process is governed by quantum tunneling. In fact, it is assumed that the mass of H atom is small enough to transmit the barrier via the quantum tunneling. Here the tunneling rate is roughly estimated by the WKB approximation.

$$R = \nu \exp\left\{ -\frac{2}{\hbar} \int_{-d/2}^{d/2} \sqrt{2m(V(x) - V_0)} dx \right\} \tag{7.1}$$

where ν, d, m, and V_0 are the vibration frequency, the tunneling path length, the mass of H atom, and the zero-point energy, respectively. $V(x)$ is the potential curve approximated by a quartic function. For an OH(OD), $h\nu$ (= $2V_0$) is calculated to be 64(45) meV by the harmonic approximation of the local minima along the [001] axis, where the degree of freedom except for the H atom is neglected and it is assumed that the hydroxyl moves circularly between the inclined geometries (± 62° from the surface normal) with the O–H bond length fixed (0.99 Å). Accordingly, the path length d is estimated to be 2.1 Å. Using these parameters, the tunneling rates are calculated to be $10^{6(2)}$ s^{-1} for an OH(OD). This value suggests tunneling in the switching process is quite plausible. The tunneling switching can be observed only for an OD within the time-resolution of STM. However, for an OH the tunneling rate is too fast and the fluctuation in the current trace is expected to be smeared out due to the limited bandwidth of preamplifier of the STM.

Although DFT calculations and the WKB modeling, as described above, is an important step to understand the tunneling dynamics of H atom, it is not enough for a quantitative description. A considerable question is the role of oxygen in the tunneling process, which is neglected above, and the difference between the classical and quantum reaction paths. Recently the detailed calculations were carried out by Davidson et al. [36]. They pointed out that the substantial displacement of oxygen atom in hydroxyl was included in the minimum-energy path, namely classical transition path, of the switching. However the tunneling is acceptable when the reaction coordinate is a direct path connecting the initial and final states with only a minimum amount of the oxygen movement, suggesting quantum tunneling causes a deviation of the reaction coordinates from the classical transition path.

The quenching of the switching of hydroxyl is theoretically predicted, which is caused by an adjacent H impurity [32]. H atom, although invisible with STM, might be present on the surface as a result of the H-atom abstraction (OH production) from a water molecule and would affect the switching of hydroxyls. However, the hydroxyl monomer is always observed as paired depressions as far as it was isolated from impurities, suggesting that the H atom is not present in the vicinity of the hydroxyl species. The produced H atom may diffuse away due to its small diffusion barrier along the trough of Cu(110) [10, 30]. However the

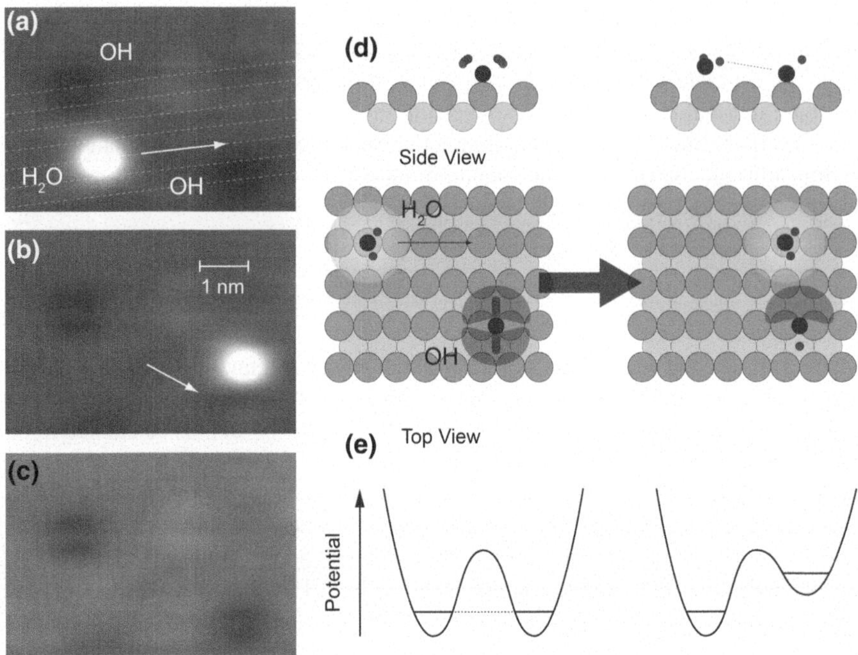

Fig. 7.7 Quenching of the tunneling switching of an OH. **a–c** Sequential STM images. **a** The water molecule is moved to the vicinity of the OH situated on the *bottom-right*. **b** The paired depression of the OH becomes single depression localized on the side of the water molecule. **c** The paired depression is resumed after the water molecule is removed. The STM images were acquired at $V_s = 24$ mV and $I_t = 0.5$ nA. **d** Schematic illustration of the quenching process. **e** Schematic potential for the hydrogen without (*left*) and with (*right*) the perturbation

quenching of the switching takes place when a water molecule comes close to a hydroxyl. Figure 7.7 shows the sequential STM images of the quenching. The water molecule is manipulated to the vicinity of the hydroxyl located on the bottom-right (Fig. 7.7a). They are separated along [001] direction by $2b_0$ (two atomic rows). After the water manipulation the paired depression of the hydroxyl is changed into the single depression localized on the side of the water molecule (Fig. 7.7b), while the other hydroxyl is still observed as a paired depression. The paired depression is recovered when the water molecule is removed (Fig. 7.7c). The expected O–O distance of $\sim$7.4 Å between the water and hydroxyl is not likely to be H-bond interaction. Thus it is assumed that the long-range interaction, such as dipole–dipole or substrate mediated interaction, works between them and consequently the potential of the switching is perturbed.

It is noted that the low-current state is substantially live longer in Fig. 7.5b. This indicates that OD prefers the orientation pointing away from the tip. It is assumed that substantial perturbation which stabilizes such an orientation exists between the tip and molecule, which is expected to be varied through changing the relative distance between the tip and molecule. Indeed, at lower current conditions,

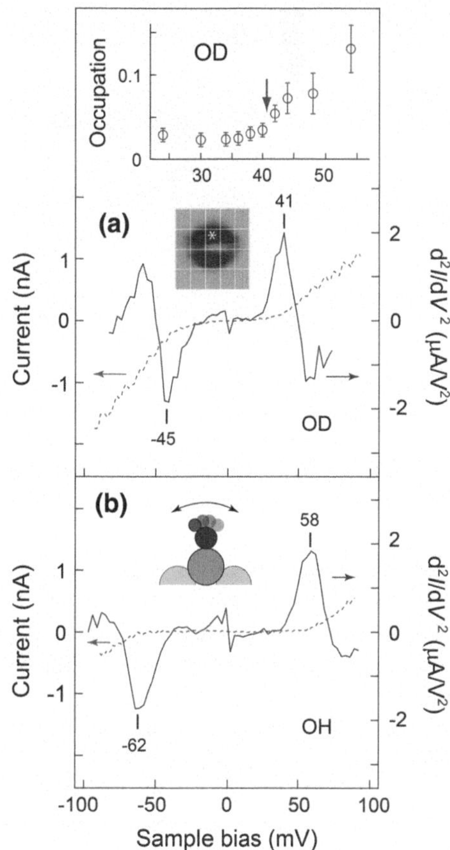

Fig. 7.8 **a** and **b** Averaged *I*–*V* curves (*dashed red curve*, left scale) and d^2I/dV^2 spectra (*solid blue curve*, right scale) measured over an OD and OH, respectively. The inset shows fractional occupation of the high-current state for an OD measured as a function of bias voltage. The tip height was fixed at $V_s = 24$ mV and $I_t = 5$ nA during the measurements. The *I*–*V* curves are average of 300 and 200 scans for OD and OH, respectively, to improve the S/N ratio. The d^2I/dV^2 spectra were measured using a lock-in amplifier with an rms modulation of 6 mV at 590 Hz

which gives a larger tip-molecule distance, the ratio between the high and low current states is changed; the ratio becomes smaller (the high current state lives longer). Although the quantitative description of this kind perturbation is still open question, the same tendency is clearly observed in Co atom hopping on Cu(111), where Co atom is also bounds to a double-well potential [17].

7.2.3 STM-IETS of a Hydroxyl Group on Cu(110)

The averaged *I*–*V* curves (red dashed lines) and the IET spectra (blue solid lines) for an OD and OH are shown in Fig. 7.8, which are measured with fixed the tip at the one depression of the STM image (indicated by asterisk in the inset image of Fig. 7.8a). The gap conditions during the measurement and the modulation voltage of the IETS are described in the caption. In order to exclude the influence of the substrate and tip, both the IET spectra and *I*–*V* curves are displayed after subtraction

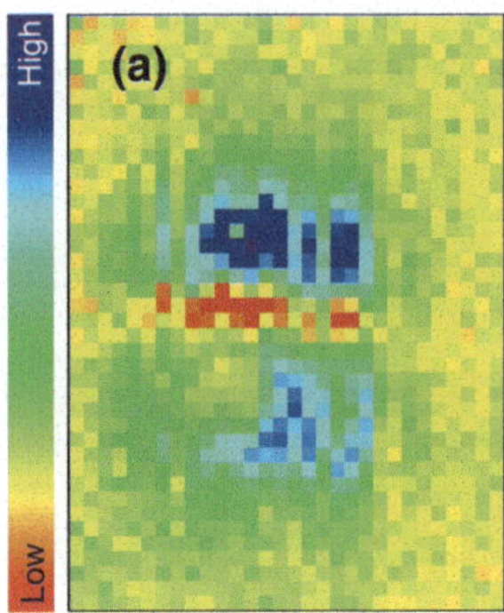

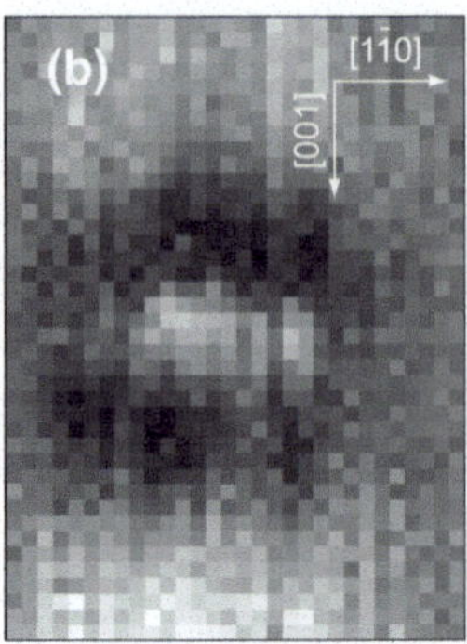

Fig. 7.9 **a** Spatial mapping of IETS intensity for an OH. At each pixel, the feedback was turned off at the tip height to give $I_t = 5$ nA at $V_s = 24$ mV, and the vibrational intensity was acquired at $V_s = 54$ mV with the modulation voltage of 12 mV. The IETS intensity ranges from 0.5 to 1.6 μA/V^2 with respect to the background. **b** Simultaneously acquired topographic image. The image size is 11×14 Å^2

with those measured over the clean surface. The spectra show peak (dip) at 41 (−45) and 58 (−62) mV for OD and OH, respectively. The averaged *I*–*V* curves are characterized by an increase and decrease at the voltages corresponding to the peak and dip in IET spectra. The peak and dip energy in IET spectra show no dependence on the relative distance between the tip and molecule. The energy and isotope ratio of ~1.4 in the peak (dip) suggests that the OH(OD) bending mode is associated with the vibration mode (inset of Fig. 7.8b). The vibrational energy of 58(41) meV for an OH(OD) is consistent with the values estimated by DFT calculations, in which the energy is estimated to be 64(45) meV using a harmonic approximation of the local minima along the [001] axis. Also the previous calculations predicted a comparable value (~60 meV) for the OH bending mode [31].

The vibrational excitation also gives rise to the change in the dynamics of hydroxyl. The switching can be enhanced by the bending mode excitation that is directly associated with the switching reaction coordinate. The quantitative analysis of the change in the current trace enables us to discuss the dynamics. The voltage dependence of the fractional occupation of the high and low current state was investigated. Here the fractional occupation $O_{\mathrm{high(low)}}$ is defined by

$$O_{\mathrm{high(low)}} = \frac{t_{\mathrm{high(low)}}}{t_{\mathrm{high}} + t_{\mathrm{low}}} \tag{7.2}$$

where $t_{\mathrm{high(low)}}$ is the time stayed in the high(low) current state during the measurement. Figure 7.8a shows the voltage dependence of the fractional occupation of the high current state. The occupation increases at the onset voltage of ~40 mV that coincides with the peak in the IETS. This result indicates that the vibrational excitation causes the enhancement of the switching and modifies the occupation of the two states. It is noted that the energy transferred from inelastic tunneling electrons, namely the vibrational energy of ~40 mV, is much lower than the barrier of 0.14 eV, suggesting the vibrationally-assisted tunneling is involved in the process.

A spatial mapping of IETS intensity enables us to visualize a spatial distribution of a specific vibrational mode with a sub-molecular resolution [37]. Figure 7.9 show the two-dimensional mapping of the IETS peak intensity measured at $V_s = 54$ mV (Fig. 7.9a) and the simultaneously acquired topographic image (Fig. 7.9b) for an OH. The intensity shows the highest over the depressions in the topographic image. Furthermore the inversion of IETS intensity can be observed at the center of molecule (between the two depressions). Either peak or dip in the IETS was observed in various adsorbate systems [38], which was explained by the coupling of the vibrational mode with the electronic structure in the tunnel junction [39]. In the present case, however, both a peak and dip coexist in the same molecule and cannot be rationalized theoretically at the moment. I speculate the enhancement of the switching induces the change of the mean distribution of H atom at the transition state, which might temporally perturb the electronic structure and eventually causes the inversion of the IETS intensity. The region of the negative intensity is very narrow along the [001] direction ($\sim$1.5 Å) and it depends on the tip conditions if the negative intensity can be observed or not.

7.3 Summary

Hydroxyl group was produced on a Cu(110) surface by the dissociation of a water molecule. The structure and dynamics were investigated by a combination of STM and DFT calculations. It was found that a hydroxyl has an inclined geometry against the surface normal and switches back and forth between the two orientations via tunneling, resulting in the characteristic paired depression aligned along the [001] axis in the STM image. This switching was directly observed only for an OD as a bi-stable fluctuating current measured over a molecule. Theoretical calculations predicted a significantly high barrier for the switching, which cannot be overcome via a thermal activation at 6 K. However, the tunneling process was acceptable because of a small mass of hydrogen. Due to its light mass the switching rate of an OH is expected to be too fast to observe with the limited time-resolution of STM. Furthermore, the switching was enhanced by the excitation of the bending mode that directly correlates with the reaction coordinate. The enhanced motion gives rise to a peak or a dip in IETS, depending on the position of the tip over the molecule.

References

1. R.P. Bell, *The Tunnel Effect in Chemistry* (Chapman and Hall, London, 1980)
2. K.W. Kehr, *Hydrogen in Metals I and II*, ed. by G. Alefeld, J. Völkl (Springer, Berlin, 1978)
3. Y. Moritomo, Y. Tokura, N. Nagaosa, T. Suzuki, K. Kumagai, Phys. Rev. Lett. **71**, 2833 (1993)

4. J.M.J. Swanson, C.M. Maupin, H. Chen, M.K. Petersen, J. Xu, Y. Wu, G.A. Voth, J. Phys. Chem. B **111**, 4300 (2007)
5. A. Kohen, R. Cannio, S. Bartolucci, J.P. Klinman, J.P. Klinman, Nature **399**, 496 (1999)
6. F. Hund, Z. Phys. **43**, 805 (1927)
7. K. Christmann, Surf. Sci. Rep. **9**, 1 (1988)
8. M.J. Puska, R.M. Nieminen, M. Manninen, B. Chakraborty, S. Holloway, J.K. Nørskov, Phys. Rev. Lett. **51**, 1081 (1983)
9. C.M. Mate, G.A. Somorjai, Phys. Rev. B **34**, 7417 (1986)
10. M. Nishijima, H. Okuyama, N. Takagi, T. Aruga, W. Brenig, Surf. Sci. Rep. **57**, 113 (2005)
11. R. Wortman, R. Gomer, R. Lundy, J. Chem. Phys. **27**, 1099 (1957)
12. R. Gomer, R. Wortman, R. Lundy, J. Chem. Phys. **27**, 1147 (1957)
13. L.J. Lauhon, W. Ho, Phys. Rev. Lett. **85**, 4566 (2000)
14. L.J. Lauhon, W. Ho, Phys. Rev. Lett. **89**, 079901(E) (2002)
15. A.J. Heinrich, C.P. Lutz, J.A. Gupta, D.M. Eigler, Science **298**, 1381 (2002)
16. J. Repp, G. Meyer, K.-H. Rieder, P. Hyldgaard, Phys. Rev. Lett. **91**, 206102 (2003)
17. J.A. Stroscio, R.J. Celotta, Science **306**, 242 (2004)
18. P.J. Feibelman, Science **295**, 99 (2002)
19. M. Forster, R. Raval, A. Hodgson, J. Carrasco, A. Michaelides, Phys. Rev. Lett. **106**, 046103 (2011)
20. M. Nagasaka, H. Kondoh, K. Amemiya, T. Ohta, Y. Iwasawa, Phys. Rev. Lett. **100**, 106101 (2008)
21. G.B. Fisher, J.L. Gland, Surf. Sci. **94**, 446 (1980)
22. G.B. Fisher, B.A. Sexton, Phys. Rev. Lett. **44**, 683 (1980)
23. T.S. Wittrig, D.E. Ibbotson, W.H. Weinberg, Surf. Sci. **102**, 506 (1981)
24. C. Ammon et al., Chem. Phys. Lett. **377**, 163 (2003)
25. G. Gilarowski, W. Erley, H. Ibach, Surf. Sci. **351**, 156 (1996)
26. L.J. Lauhon, W. Ho, J. Phys. Chem. **105**, 3987 (2001)
27. K. Morgenstern, K.-H. Rieder, Chem. Phys. Lett. **358**, 250 (2002)
28. Mugarza, T. K. Shimizu, D F. Ogletree, M. Salmeron, Surf. Sci. **603**, 2030 (2009)
29. H.-J. Shin, J. Jung, K. Motobayashi, S. Yanagisawa, Y. Morikawa, Y. Kim, M. Kawai, Nat. Mater. **9**, 442 (2010)
30. Q.-L. Tang, Z.-X. Chen, J. Chem. Phys. **127**, 104707 (2007)
31. J. Ren, S. Meng, J. Am. Chem. Soc. **128**, 9282 (2006)
32. J. Ren, S. Meng, Phys. Rev. B **77**, 054110 (2008)
33. B.G. Briner, M. Doering, H.-P. Rust, A.M. Bradshaw, Phys. Rev. Lett. **78**, 1516 (1997)
34. P. Avouris, R.E. Walkup, A.R. Rossi, H.C. Akpati, P. Nordlander, T.-C. Shen, G.C. Abeln, J.W. Lyding, Surf. Sci. **363**, 368 (1996)
35. M. Polak, Surf. Sci. **321**, 249 (1994)
36. E.R.M. Davidson, A. Alavi, A. Michaelides, Phys. Rev. B **81**, 153410 (2010)
37. B.C. Stipe, M.A. Rezaei, W. Ho, Phys. Rev. Lett. **82**, 1724 (1999)
38. W. Ho, J. Chem. Phys. **117**, 11033 (2002)
39. H. Ueba, T. Mii, S.G. Tikhodeev, Surf. Sci. **601**, 5220 (2007)

Chapter 8
Hydroxyl Dimer: Non-linear *I*–*V* Characteristics in an STM Junction

Abstract I describe the assembly and characterization of a hydroxyl dimer on a Cu(110) surface in this chapter. A dimer is produced by the reaction between a water and atomic oxygen with the STM manipulation. Hydroxyl groups in a dimer have an inclined geometry in common with a monomer and flips back and forth between two states. Although the tunneling switching observed for a monomer is quenched for a dimer due to the formation of H bond between hydroxyl groups as well as the increased mass effect, the switching can be induced by the vibrational excitation via the inelastic electron tunneling process. It is found that the switching results in non-linear characteristics in the averaged *I*–*V* curve measured over a dimer. I propose a model describing the relation between the vibrational excitation and non-linear *I*–*V* characteristics in the STM junction.

Keywords Hydroxyl groups on metal surfaces · Molecular switch · Non-linear *I*–*V* characteristics

8.1 Introduction

Tunneling dynamics is strongly affected by surrounding environments. In order to examine this kind of effects, I assembled a hydroxyl dimer on a Cu(110) surface. It is found that H-bond formation between the hydroxyl groups suppresses tunneling switching that is observed in a monomer. However, the switching between two orientations can be induced by vibrational excitations. Controlling the switching property of single molecule is a key challenge in the development of molecular based devises as well as the fabrication at the scale. Several kinds of single molecule switches have been investigated using scanning tunneling microscope (STM) [1–8] in the past decade. The switching properties rely on the change of a conformation, electronic and spin state of molecule. Furthermore the

T. Kumagai, *Visualization of Hydrogen-Bond Dynamics*, Springer Theses,
DOI: 10.1007/978-4-431-54156-1_8,

strong coupling of vibration with molecular switching eventually influences the electron transport property.

Electron transport through single molecule junctions has been receiving enthusiastic interest in last few decades because next generation electronics may rely on single molecules performing as the smallest unit of functions in electronic devices [9]. The determination of molecular conductivity at the single molecule level is a challenging subject. Electron transportation properties of nano-scale junctions including molecules can be affected by internal molecular configurations and environmental factors. In order to investigate electron transport within a single-molecule junction several experimental methods have been established, including break junction [10], mercury column method [11–14], nanowires method [15], nanolithographcally defined pores [16], capillary molecular junction [17], STM-based techniques [18–26], cross-wire method [27], metallic nanoparticle-based contact [28], and junctions prepared by the electromigration effect [29]. It is known that vibrational excitations in molecular junction results in characteristic features in the I–V curve. For instance, non-linear I–V characteristics associated with the vibrationally mediated configuration change of a molecule have been observed for a pyrrolidine on a Cu(001) [30], H_2 on Cu [31], CO bridging a Pt contact [32], and H_2 in Au contacts [33]. In these systems dI/dV spectra show anomalous spikes (not like steps usually observed in inelastic electron tunneling spectroscopy [34]) at the bias voltage related to the vibration energies. The identification of molecular vibrations makes it possible to discuss the structure of the junction. At the same time, it might be used to control the charge transport properties in molecular junctions, which is a key to the development of molecular devices.

8.2 Results and Discussions

8.2.1 Production of a Hydroxyl Dimer on Cu(110)

A hydroxyl dimer can be produced by the reaction a water molecule with atomic oxygen on Cu(110); $H_2O + O \rightarrow (OH)_2$. Figure 8.1a, b show sequential images and schematic illustrations before and after the reaction. A water molecule is located on the adjacent atomic row of an oxygen atom located at a fourfold hollow site (the geometry is illustrated in Fig. 8.1c). The water molecule is dragged along the atomic row by the STM manipulation and reacted with the oxygen atom and then yielded a hydroxyl dimer. The reaction occurs spontaneously when the reactants come close to each other. A hydroxyl dimer is observed as a semi-circular depression. On the other hand, when a water molecule is located on the row away from by $2.5b_0$, the reaction does not occur and it is weakly interacted with oxygen atom (Fig. 8.2). Although a spontaneous switching cannot be observed for a dimer, it can be switched between two orientations with a voltage pulse of the STM (Fig. 8.3).

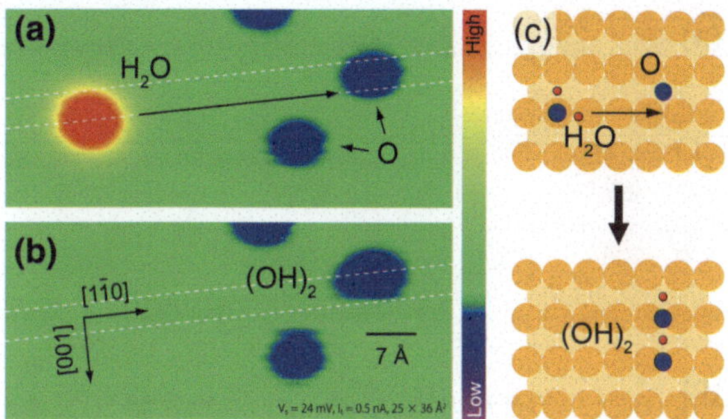

Fig. 8.1 Production of a hydroxyl dimer. **a** and **b** STM images before and after the reaction between an H_2O and O on a Cu(110) surface. A hydroxyl dimer was formed by dragging the H_2O to O. *White broken lines* represent the atomic rows of Cu(110). The STM images were acquired at $V_s = 30$ mV and $I_t = 0.5$ nA. **c** Schematic illustration of assembly process

8.2.2 Structure and Dynamics of a Hydroxyl Dimer on Cu(110)

The structure of a hydroxyl dimer was determined by DFT calculations.[1] Figure 8.4 shows the optimized structure where the each hydroxyl is bound to the short-bridge site with having an inclined geometry along the [001] direction. The dimer is found to be stabilized by 0.22 eV compared to two isolated hydroxyls due to the formation of H-bond, which is considered to be the origin of quenching the tunneling switching as well as the mass effect.

The reversible switching can be seen by recording tunneling current over a dimer. As shown in the inset of Fig. 8.5a, the switching causes a telegraph noise in the current trace. Using the same method described in the previous chapters, the switching rate was obtained. Figure 8.5 shows the voltage dependence of the switching yield. The yield $Y(V)$ (reaction probability per electron) is given by

$$Y(V) = \frac{R(V)}{I_t(V)/e} \tag{8.3}$$

where e is the elementary electric charge. During the measurement the STM tip is fixed over the depression corresponding to the orientation in which the hydrogen pointing away from the tip. This geometry gives the low current state and the tip is assumed to be located approximately over the H-bond donor molecule. The threshold voltage at which the switching is induced is ~200 mV for both of $(OH)_2$ and $(OD)_2$. After an abrupt increase around 200 mV, the yield shows a moderate and rapid increases at ~390 and ~290 mV for $(OH)_2$ and $(OD)_2$, respectively. The isotope ratio of the second increase is ~1.35, suggesting that the enhancement

[1] 1 The OH dimer is put on one side of a three-layer Cu slab arrayed in a 3 × 3 surface unit cell. The other conditions are the same with the monomer described in Sect. 7.2.2.

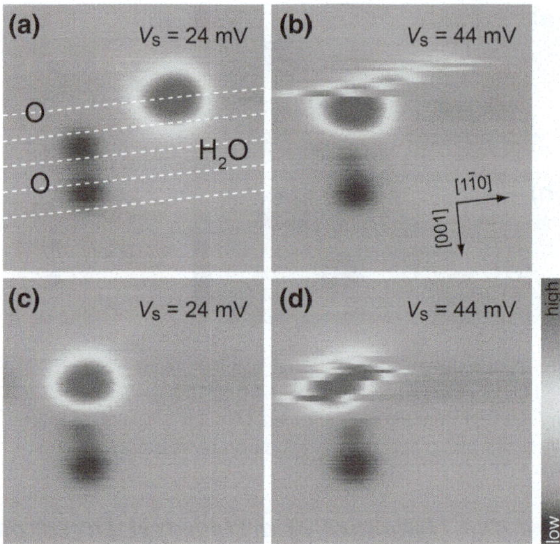

Fig. 8.2 **a** STM images of a water molecule and two oxygen atoms on a Cu(110) surface. The water molecule is located on the atomic row separated by $1.5b_0$ from the upper oxygen. The *white dashed lines* represent the atomic rows. **b** The water molecule was brought close to the oxygen by scanning the area at $V_s = 44$ mV. **c** and **d** While the appearance of the water molecule does not significantly change, the water molecule is trapped near the oxygen atom, indicating that it is weakly interacted with oxygen. All images were obtained at $I_t = 0.5$ nA and the bias voltage is shown in each image. The image size is 41×41 Å^2

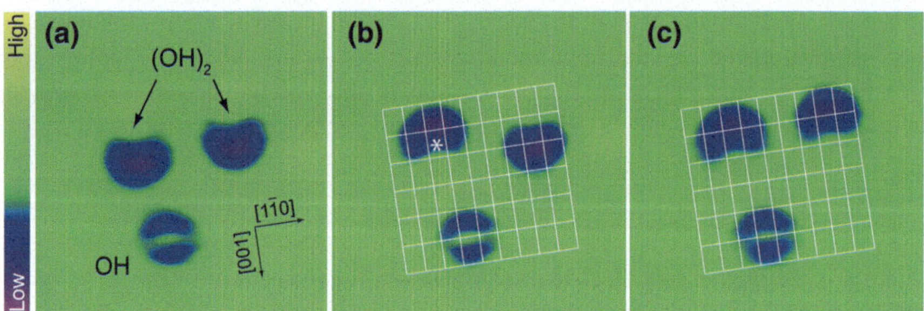

Fig. 8.3 **a** Typical images of OH dimers compared with an OH monomer. **b** One of the dimers is flipped by a voltage pulse of the STM. The *white grid lines* and the *asterisk* represent the lattice of Cu(110) and the inversion center of the flip. **c** Another one is also flipped in the same way. All images were obtained at $V_s = 24$ mV and $I_t = 5$ nA, and size is 40×40 Å^2

is associated with the O–H(D) stretch excitation. It is noted that the vibrational energy is significantly red-shifted from that of free OH (443 meV [35]). The energy of ν(OH) is directly related to the strength of H bond [36]. In general, H-bond formation causes the weakening of covalent nature of O–H bond via a charge-transfer, eventually giving rise to significant red-shift. Therefore the result

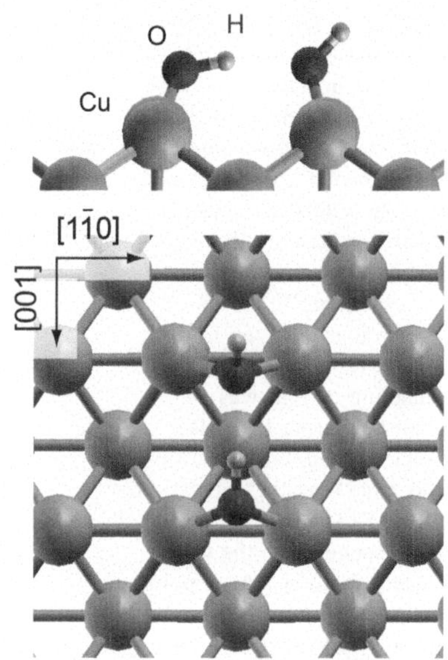

Fig. 8.4 Side (*upper*) and top (*lower*) views of the calculated structure for an OH dimer on Cu(110). The H-bond OH-O distance is 1.83 Å. The OH bond length and the tilt angles to the surface normal for H-bond donor (acceptor) are 1.00 (0.98) Å and 78.6° (43.4°), respectively

obviously indicates that H-bonding interaction exists in a dimer. For H-bonded OH groups on Pd(100), the OH stretch mode was observed at 403 meV [37].

On the other hand, Fig. 8.5b shows the current dependence of the switching rate. The rate shows a linear dependence at high bias voltages, but it changes into second-order dependence at low bias. Thus the switching is induced by one- and two-electron processes at the high and low bias voltages, respectively. As shown in the inset of Fig. 8.5b the crossover from one- to two-electron processes takes place around $V_s \sim 200$ mV. The mechanism of the first increase around 200 mV is still unclear. There is no specific vibration modetrun 1698around this energy region for either of $(OH)_2$ and $(OD)_2$. Furthermore, a significant isotope effect is also observed in the switching rate; the rate of $(OD)_2$ is smaller than that of $(OH)_2$ by two orders of magnitude. Although this origin is also still open question, I speculate the tunneling at the excited state might be included in the process.

8.2.3 Non-linear Characteristics of the Averaged I–V Curve of a Hydroxyl Dimer

Figure 8.6 shows the averaged *I*–*V* curve and the conductance (d*I*/d*V*) spectra measured over an $(OD)_2$. The STM tip was fixed over the depression at a gap corresponding to $V_s = 24$ mV and $I_t = 5$ nA before opening the feedback loop. The *I*–*V* curve (dashed line) exhibits a rapid increase at the energy of ν(OD). This non-linear character is observed as a sharp peak in the d*I*/d*V* spectra (solid lines). The same feature can be observed at a negative bias (inset of Fig. 8.6) as well as for an $(OH)_2$. It is clear that the molecular vibration associated with non-linear

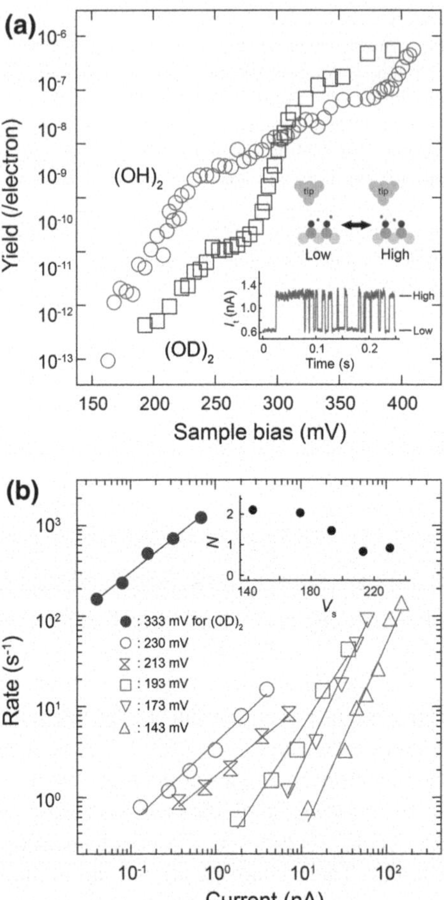

Fig. 8.5 **a** The voltage dependence of the switching yield for an $(OH)_2$ (*red circles*) and $(OD)_2$ (*blue squares*). The *upper-left* inset is the current trace measured over an $(OD)_2$. The current fluctuates between the high and low current states corresponding to the orientations depicted in the *bottom-right* inset. The tunneling current was recorded during the pulse voltage of 303 mV. The tip was fixed over the depression of the STM image (corresponding to the low current state) at the set point of $I_t = 0.05$ nA and $V_s = 24$ mV, and then the feedback was turned off and the voltage was applied. The yields were measured with the tip fixed at the low current state. **b** The current dependence of switching rate. The inset shows voltage dependence of the reaction order for an $(OH)_2$

characteristics in the I–V curve. To elucidate the mechanism, the voltage dependence of the fractional occupations of the two states was investigated. The tunneling current ($I_{\text{high, low}}$) and the fractional occupation ($O_{\text{high, low}}$) at the high and low states are obtained as a function of the voltage (Fig. 8.7a). The fractional occupation is defined as the ratio of the time spent at which the high and low current states in the current trace. The tunneling current at the high and low states shows a linear increase (bottom of Fig. 8.7a). On the other hand, the fractional occupations of each state show an abrupt change at the threshold energy of the ν(OD) excitation (upper of Fig. 8.7a). The average current I_{average} at a given bias can be reproduced from the following equation

$$I_{\text{average}} = O_{\text{high}} \times I_{\text{high}} + O_{\text{low}} \times I_{\text{low}} \tag{8.4}$$

I_{average} is plotted in Fig. 8.7b, which coincides with the averaged I–V curve. It is now clear that the non-linear characteristics in the averaged I–V curves are

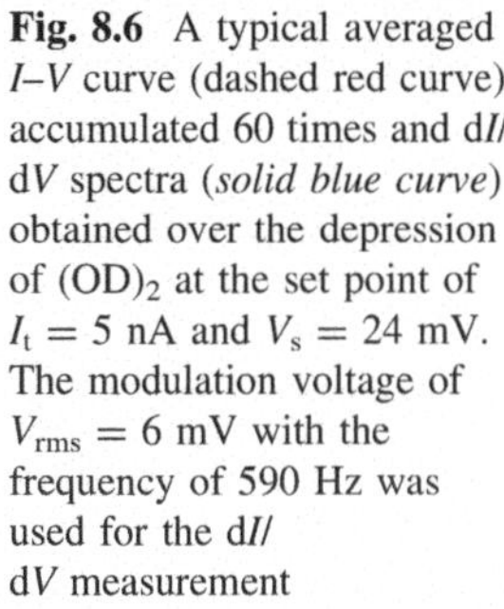

Fig. 8.6 A typical averaged *I–V* curve (dashed red curve) accumulated 60 times and d*I*/d*V* spectra (*solid blue curve*) obtained over the depression of $(OD)_2$ at the set point of $I_t = 5$ nA and $V_s = 24$ mV. The modulation voltage of $V_{rms} = 6$ mV with the frequency of 590 Hz was used for the d*I*/d*V* measurement

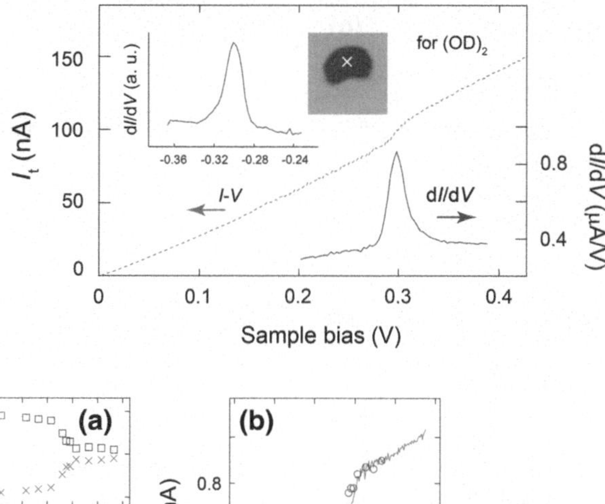

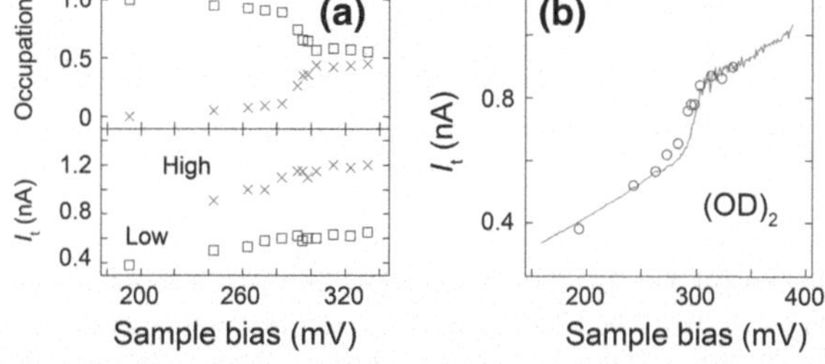

Fig. 8.7 **a** The fractional occupation (*upper*) and the state-resolved *I–V* plot (*lower*) for the low (*blue squares*) and high (*red crosses*) current states as a function of the applied bias voltage measured for $(OD)_2$. The *I–V* plot reproduced by using these data is shown in **b** (*blue circles*) with an averaged *I–V* curve (*red line*). These data were obtained at the set point of $I_t = 0.05$ nA and $V_s = 24$ mV

attributed to the dynamical property of a dimer, that is, the change in the fractional occupation of the two states caused by the ν(OH(D)) excitation. Figure 8.8 shows a schematic illustration of the mechanism. At the initial geometry the STM tip is fixed over the depression of a dimer that corresponds to the low current state (Fig. 8.8a). Before the vibrational excitation, the occupation of the low current state is dominant (Fig. 8.8c). After the excitation, the enhancement of the switching causes a significant change in the occupation; the low current and high current states are decreased and increased, respectively (Fig. 8.8d), eventually giving rise to the increase of the average current. Furthermore, the d^2I/dV^2 spectra (Fig. 8.9) are characterized by the peak and the subsequent dip which correspond to the onset and saturation of the occupation changes, respectively.

8.3 Summary

A hydroxyl dimer on Cu(110) is produced by the reaction of a water molecule with an oxygen atom. Hydroxyl groups adsorb on the twofold short-bridge site and is titled along the [001] direction in common with the monomer. The switching

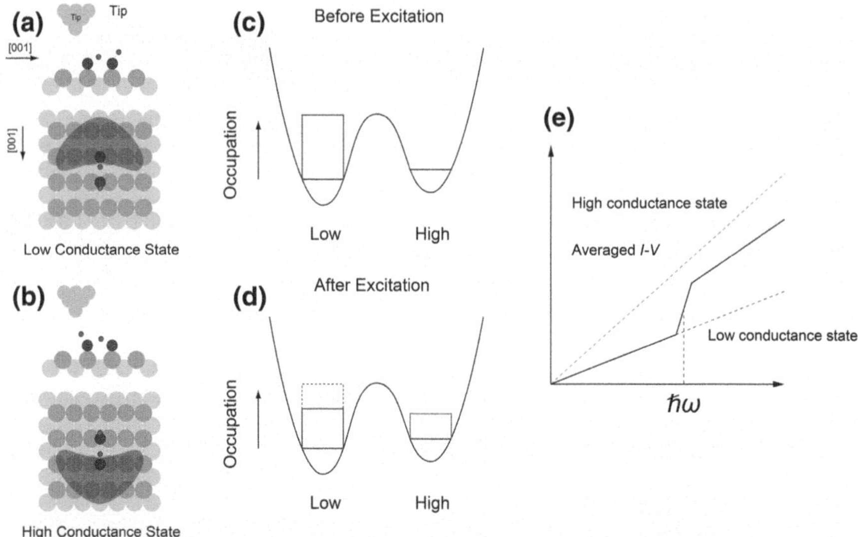

Fig. 8.8 Schematic model describing the mechanism for the non-linear *I–V* characteristics of a hydroxyl dimer on Cu(110). **a** The STM tip is fixed over the depression (low conductance state) before ramping up the bias voltage. **b** When the dimer is flipped, the appearance under the tip becomes brighter than (**a**). (high conductance state). **c** Before the vibrational excitation the occupation of the low conductance state is dominant. **d** After the excitation the switching is dramatically enhanced, resulting in the reduction (increase) of the occupation of the high (low) conductance state. **e** In this model the stepwise change is predicted at the vibration energy

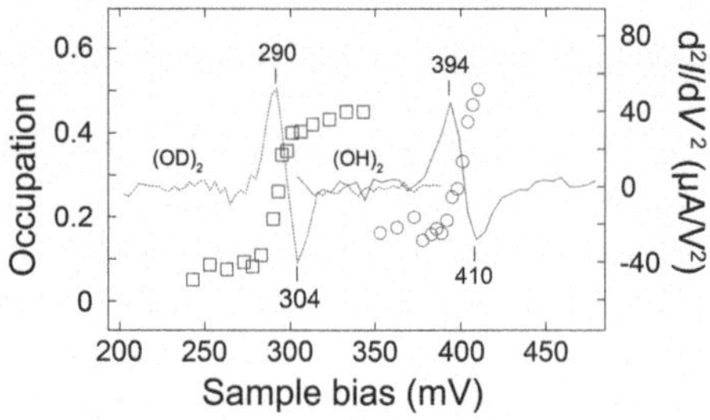

Fig. 8.9 The IETS (d^2I/dV^2) spectra (*solid lines*) and the fractional occupations for an $(OD)_2$ (*open squares*) and $(OH)_2$ (*open circles*). The peak and dip voltages in the spectra were determined by fitting the data to the derivative of Gaussian function. The spectra were obtained with the STM tip fixed over the depression at the set point of $I_t = 0.5$ and 5 nA at $V_s = 24$ mV for an $(OH)_2$ and $(OD)_2$, respectively. An $(OH)_2$ would be broken during the measurement with the set point of 5 nA. The modulation voltage of $V_{rms} = 6$ mV with the frequency of 590 Hz was used for d^2I/dV^2 measurement. The fractional occupations were also measured with the STM tip fixed over the depression at the set point of $I_t = 0.05$ and $V_s = 24$ mV for an $(OH)_2$ and $(OD)_2$

between two orientations can be induced by the vibrational excitation, while tunneling process is quenched because of the H-bond formation as well as the mass effect. The drastic enhancement of the switching can be observed at the energy of the O–H(D) stretch excitation. This strong coupling between the vibration and switching results in the non-linearity in the averaged *I*–*V* characteristics measured over the dimer, which manifests a peak in the d*I*/d*V* spectra.

References

1. F. Moresco, G. Meyer, K-H. Rieder, H. Tang, A. Gourdon, C. Joachim, Phys. Rev. Lett. **86**, 672 (2001)
2. S.W. Wu, N. Ogawa, W. Ho, Science **312**, 1362 (2006)
3. V. Iancu, S.-W. Hla, Proc. Natl. Acad. Sci. USA **103**, 13718 (2006)
4. J. Henzl, M. Mehlhorn, H. Gawronski, K-H. Rieder, K. Morgenstern, Chem. Int. Ed. **45**, 603 (2006)
5. B.-Y. Choi, S.-J. Kahng, S. Kim, H. Kim, H.W. Kim, Y.J. Song, J. Ihm, Y. Kuk, Phys. Rev. Lett. **96**, 156106 (2006)
6. M.J. Comstock, N. Levy, A. Kirakosian, J. Cho, F. Lauterwasser, J.H. Harvey, D.A. Strubbe, J.M.J. Fréchet, D. Trauner, S.G. Louie, M.F. Crommie, Phys. Rev. Lett. **99**, 038301 (2007)
7. Y. Wang, J. Kröger, R. Berndt, W.A. Hofer, J. Am. Chem. Soc. **131**, 3639 (2009)
8. T. Komeda, H. Isshiki, J. Liu, Y-F. Zhang, N. Lorente, K. Katoh, B.K. Breedlove, M. Yamashita, Nate commun **2**, 217 (2011)
9. A. Aviram, M. Ratner, *Molecular Electronics: Science and Technology* (New York Academy of Sciences, New York, 1998)
10. M.A. Reed, C. Zhou, C.J. Muller, T.P. Burgin, J.M. Tour, Science **278**, 252 (1997)
11. A. Rampi, O.J.A. Schueller, G.M. Whitesides, Appl. Phys. Lett. **72**, 1781 (1998)
12. R. Haag, M.A. Rampi, R.E. Holmlin, G.M. Whitesides, J. Am. Chem. Soc. **121**, 7895 (1999)
13. R.E. Holmlin, R. Haag, M.L. Chabinyc, R.F. Ismagilov, A.E. Cohen, A. Terfort, M.A. Rampi, G.M. Whitesides, J. Am. Chem. Soc. **123**, 5075 (2001)
14. M.A. Rampi, G.M. Whitesides, Chem. Phys. **281**, 373 (2002)
15. J.K.N. Mbindyo, T.E. Mallouk, J.B. Mattzela, I. Kratochvilova, B. Razavi, T.N. Jackson, T.S. Mayer, J. Am. Chem. Soc. **124**, 4020 (2002)
16. C. Zhou, M.R. Deshpande, M.A. Reed, L. Jones, J.M. Tour, Appl. Phys. Lett. **71**, 611 (1997)
17. X.-L. Fan, C. Wang, D.-L. Yang, L.-J. Wan, C. Bai, Chem. Phys. Lett. **361**, 465 (2002)
18. L.A. Bumm, J.J. Arnold, M.T. Cygan, T.D. Dunbar, T.P. Burgin, L. Jones II, D.L. Allara, J.M. Tour, P.S. Weiss, Science **271**, 1705 (1996)
19. L.A. Bumm, J.J. Arnold, T.D. Dunbar, D.L. Allara, P.S. Weiss, J. Phys. Chem. B **103**, 8122 (1999)
20. S. Datta, W.D. Tian, S.H. Hong, R. Reifenberger, J.I. Henderson, C.P. Kubiak, Phys. Rev. Lett. **79**, 2530 (1997)
21. Y.Q. Xue, S. Datta, S.H. Hong, R. Reifenberger, J.I. Henderson, C.P. Kubiak, Phys. Rev. B **59**, R7852 (1999)
22. C. Zeng, H. Wang, B. Wang, J. Yang, J.G. Hou, Appl. Phys. Lett. **77**, 3595 (2000)
23. K. Takanashi, S. Mitani, J. Chiba, H. Fujimori, J. Appl. Phys. **87**, 6331 (2000)
24. X.D. Cui, A. Primak, X. Zarate, J. Tomfohr, O.F. Sankey, A.L. Moore, T.A. Moore, D. Gust, G. Harris, S.M. Lindsay, Science **264**, 571 (2001)
25. F.-R.F. Fan, J. Yang, L. Cai, D.W. Price Jr, S.M. Dirk, D.V. Kosynkin, Y. Yao, A.M. Rawlett, J.M. Tour, A.J. Bard, J. Am. Chem. Soc. **124**, 5550 (2002)
26. D.J. Wold, R. Haag, M.A. Rampi, C.D. Frisbie, J. Phys. Chem. B **106**, 2813 (2002)
27. J.G. Kushmerick, D.B. Holt, S.K. Pollack, M.A. Ratner, J.C. Yang, T.L. Schull, J. Naciri, M.H. Moore, R. Shashidhar, J. Am. Chem. Soc. **124**, 10654 (2002)

28. M. Dorogi, J. Gomez, R. Osifchin, R.P. Andres, R. Reifenberger, Phys. Rev. B **52**, 9071 (1995)
29. H. Park, A.K.L. Lim, A.P. Alivisatos, J. Park, P.L. McEuen, Appl. Phys. Lett. **75**, 301 (1999)
30. J. Gaudioso, L.J. Lauhon, W. Ho, Phys. Rev. Lett. **85**, 1918 (2000)
31. J.A. Gupta, C.P. Lutz, A.J. Heinrich, D.M. Eigler, Phys. Rev. B **71**, 115416 (2005)
32. W.H.A. Thijssen, D. Djukic, A.F. Otte, R.H. Bremmer, J.M. van Ruitenbeek, Phys. Rev. Lett. **97**, 226806 (2006)
33. A. Halbritter, P. Makk, Sz. Csonka, and G. Mihály. Phys. Rev. B **77**, 075402 (2008)
34. B.C. Stipe, M.A. Rezaei, W. Ho, Science **280**, 1732 (1998)
35. L.E. Firment, G.A. Somorjai, J. Chem. Phys. **63**, 1037 (1975)
36. D.A. Schmidt, K. Miki, J. Phy, Chem. A **111**, 10119 (2007)
37. M. Stuve, S.W. Jorgensen, R.J. Madix, Surf. Sci. **146**, 179 (1984)

Chapter 9
Water-Hydroxyl Complexes: Direct Observation of a Symmetric Hydrogen Bond

Abstract I describe the hydration of a hydroxyl group with a water molecule on a Cu(110) surface and the characterization of water-hydroxyl complexes in this chapter. Two different structural isomers are selectively produced depending on the initial geometry of the reactants before the reaction. These isomers are employed as a model system to examine the nature of H-bond. A combination of STM experiments with DFT calculations reveals that one of the isomers forms "a low-barrier H bond" due to the strong interaction between water and hydroxyl, where the zero-point nuclear motion plays a crucial role to determine the structure.

Keywords Hydration reaction · Water-hydroxyl complex · Low-barrier hydrogen bond

9.1 Introduction

Nature of shared H atom/proton in H bond is a key interest to understand H/proton-transfer reactions, which plays important roles in the elementary process of chemical and biological reactions [1–4]. It is well-known that proton shows an anomalously high mobility in liquid water [5]. In order to explain this anomaly Grotthuss proposed the idea of "structural diffusion" in 1806 [6]. In the Grotthuss mechanism proton transfer is described by a sequential exchange of H- and covalent-bonds between water molecules as shown in Fig. 9.1. This concept has been refined by invoking thermal hopping, proton tunneling, and solvation effects. The Zundel and Eigen cations (Fig. 9.2) were proposed to be plausible hydrated structures associating with the transfer process. The strength and symmetry of the H bond in those cations are primary issues. In general, the strength of H bond is reflected in the distance between two O atoms bridged by H atom [7, 8]. The O–O distance ($d_{O\text{-}O}$) determines the potential energy surface (PES) of a shared H/proton

T. Kumagai, *Visualization of Hydrogen-Bond Dynamics*, Springer Theses,
DOI: 10.1007/978-4-431-54156-1_9, © Springer Japan 2012

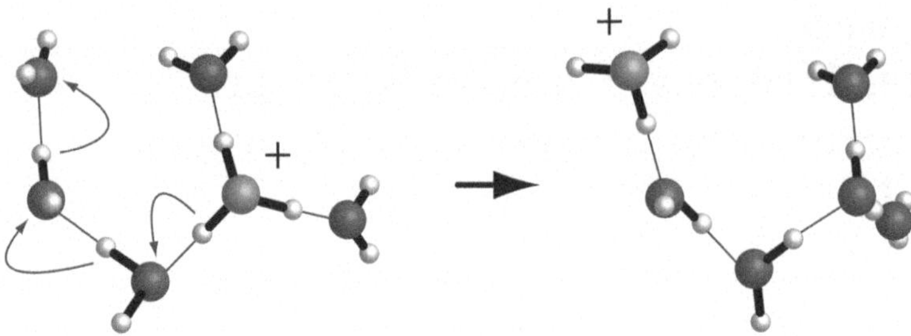

Fig. 9.1 Grotthuss mechanism. "Structural diffusion" of an excess proton through the H-bond network of water molecules. The process includes the sequential cleavage and formation of covalent bond

and the strength can be classified by $d_{O\text{-}O}$. For normal H bond, the $d_{O\text{-}O}$ is $\sim$2.8 Å and the double-well potential with a relatively high barrier is expected, where the H atom is covalently bonded to one of the O atoms (Fig. 9.3a). The ground state of the shared H is much below than the barrier and the wave packet is localized in one well. In this case H transfer takes place via the over-barrier or tunneling process. On the other hand, when the $d_{O\text{-}O}$ becomes $\sim$2.5 Å, the H transfer barrier is significantly reduced and the zero-point nuclear motion plays a decisive role (Fig. 9.3b). In this situation the zero-point energy (ZPE) overcomes the reduced barrier and the probability amplitude of the shared H/proton is localized around the center of two O atoms. This type of H bond is called "a low barrier H bond" or "a symmetric H bond", where the H transfer via tunneling is negligible and the simple transition-state theory no longer works. The smaller $d_{O\text{-}O}$, less than 2.5 Å, results in a single-well potential and a very strong H bond. The strength of H bonds is characterized by a significant red-shift of ν(OH).

H-bonded complexes $H_5O_2^+$ and $H_3O_2^-$ are considered as a prototype to examine such a strong H bonding. Tuckerman et al. highlighted the impact of quantum nature of the shared proton in $H_5O_2^+$ and $H_3O_2^-$ [3]. They investigated the relative influence of thermal and quantum fluctuations on the proton transfer with ab initio path-integral molecular dynamics (PIMD). In PIMD, both the electrons and nuclei are treated as quantum particles in contrast to traditional molecular dynamics in which the nuclei are treated as classical point-like particles (only electrons are treated as quantum particles). The PIMD calculations predicted that no barrier exists for the shared proton in $H_5O_2^+$ and results in a strong H bond, while $H_3O_2^-$ forms "a low-barrier H bond" and the ZPE plays a crucial role. Such a symmetric H bond is actually observed in the x-ray diffraction of a compressed ice, wherein the $d_{O\text{-}O}$ reached 2.4 Å under $\sim$ 60 GPa [9]. Additionally, isotope substitution of the shared H into deuterium has a significant impact on the strength of H-bond because of the influence of the mass on tunneling rates, the ZPE, and the motional properties. Indeed, many isotope effects are found in kinetics and even for geometrical parameters of H-bonded systems; the latter is known as the Ubbelohde effect [10]. Although this kind of structural isotope effect virtually

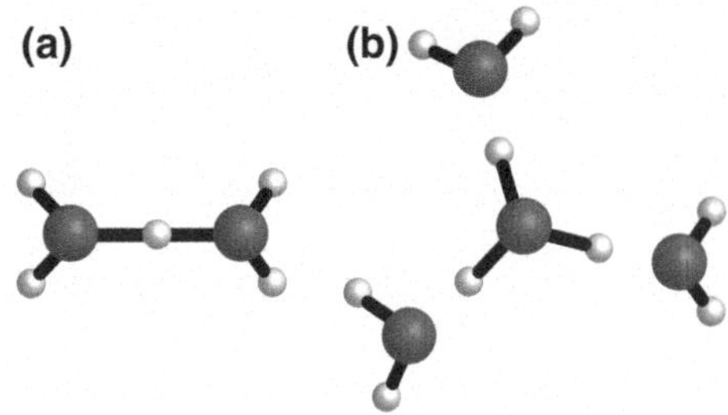

Fig. 9.2 Hydration structure of proton in solvent. (a) Zundel ($H_5O_2^+$) and (b) Eigen cation ($H_9O_4^+$)

vanishes for crystalline water ice under normal conditions, it was observed in water bilayer on Ru(001) [11].

In surface chemistry, water-hydroxyl complexes have been particular key species in the process of wetting, corrosion, and electrode reaction. Both of experiments and theoretical calculations predicted the first contact layer of water is not consisted of pure water phase but of a mixture of water and hydroxyl in several cases [12–21]. Water-hydroxyl overlayers have been observed on close-packed metal surfaces in the past. It was unveiled that the existence of hydroxyls leads to stabilization of water adsorption. The structure of such mixed-overlayers is determined by the competition between the molecule–substrate and water-hydroxyl interactions as well as intact water layers. The distance between the adjacent molecules is strongly affected by the substrate geometry. For instance, the average distance between the molecules is estimated to be ~2.83 Å on Pt(111) and ~2.50 Å on Ni(111). In the latter case the $d_{O\text{-}O}$ is already in the region of a low-barrier H bond. Recently, PIMD was performed for the water-hydroxyl wetting layer on the Pt(111), Ru(0001) and Ni(111) surfaces [22]. It is revealed that the partial and entirely symmetric delocalization of the shared H occurs on Pt (Ru) and Ni surfaces, respectively, highlighting the effect of metal substrates serving to reduce the classical H transfer barriers within the overlayers. However the water-hydroxyl interaction, especially its local H-bonding nature, is still unknown at the molecule level.

9.2 Results and Discussions

9.2.1 Production of Water-Hydroxyl Complexes on Cu(110)

The hydration between a hydroxyl group and a water molecule can be induced by the STM manipulation. An isolated hydroxyl group is produced by the dissociation of a water molecule. Figures 9.4a and b show the sequential images of the reaction (Fig. 9.4c is a schematic illustration of the reaction). The black dashed line in the image represents the atomic row of Cu(110). The hydroxyl group and water monomer are adsorbed on a twofold short-bridge and on-top site, respectively. A water and hydroxyl locating on the same atomic row are reacted in Fig. 9.4 a–b

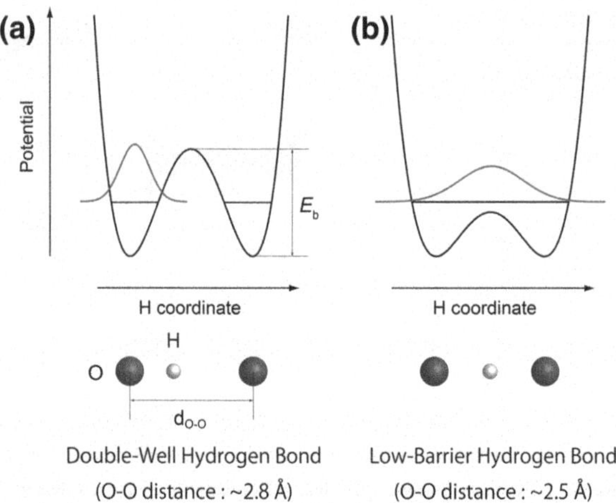

Fig. 9.3 Schematic energy diagram of a shared H/proton. **a** In a normal H bond the shared H/proton is in a double well potential and the ZPE is well-below the barrier height E_b. The wave packet is localized in the well and the H-transfer is described by over-barrier (classical) or tunneling (quantum) process. **b** The compression of the $d_{O\text{-}O}$ causes the significant reduction of the E_b and the ZPE eventually exceeds the E_b. The H-transfer via tunneling is negligible and the simple transition-state theory no longer works

and yield a pear-shaped protrusion (label this product as $[1\bar{1}0]$-complex). The reaction occurs spontaneously when the reactants sufficiently come close to each other. A product complex is never dissociated into the parent water and hydroxyl under low bias conditions, suggesting it is sufficiently stable compared to an isolated water and hydroxyl. A $[1\bar{1}0]$-complex can be flipped between two orientations with a voltage pulse of STM (Fig. 9.4d–e). The thresholds of the flipping are roughly estimated to be $\sim$ 70 and $\sim$170 mV for H- and D-complexes, respectively. As a higher bias voltage ($\sim$200 mV) is applied, a $[1\bar{1}0]$-complex dissociates into a water molecule and hydroxyl group. Thus, we can control the formation and dissociation of H bond in reversible fashion at the single molecule precision. Figures 9.5a–d show the sequential images of the formation and dissociation of the complex. The flip motion and dissociation can be directly monitored by recording tunneling current during a voltage pulse with the feedback turned off. Figure 9.5(e) shows a typical result measured over an H_2O-OH $[1\bar{1}0]$-complex at $V_s = 179$ mV. The flipping gives relatively wide feature in the high current state and the abrupt drops in the current correspond to the moment of the dissociation. In Fig. 9.5e, the dissociation occurs twice and the complex was reformed after the first dissociation and finally dissociated into a water and hydroxyl. The dissociation rate was estimated by repeating the measurement and the threshold of the dissociation was determined to be $\sim$180 mV, indicating the water scissor mode is associated with the dissociation process. For a D_2O-OD $[1\bar{1}0]$-complex, the threshold was $\sim$220 mV, suggesting the overtone excitation of the D_2O scissors mode.

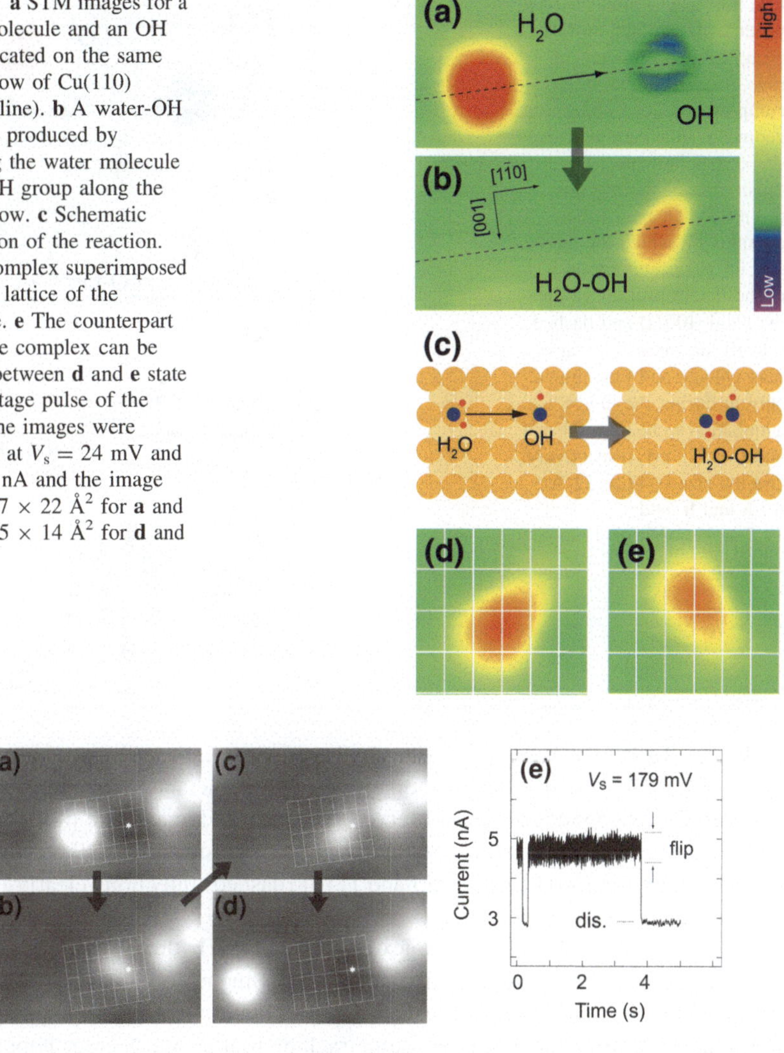

Fig. 9.4 **a** STM images for a water molecule and an OH group located on the same atomic row of Cu(110) (dashed line). **b** A water-OH complex produced by dragging the water molecule to the OH group along the atomic row. **c** Schematic illustration of the reaction. **d** The complex superimposed with the lattice of the substrate. **e** The counterpart of **d**. The complex can be flipped between **d** and **e** state by a voltage pulse of the STM. The images were acquired at $V_s = 24$ mV and $I_t = 0.5$ nA and the image size is 47×22 Å^2 for **a** and **b**, and 15×14 Å^2 for **d** and **e**

Fig. 9.5 Sequential STM images of a water-hydroxyl complex formation **a–b**, flipping **b–d** and dissociation **c–d**. The grid lines represent the lattice of Cu(110) and the yellow stars represent the O position of hydroxyl. **e** A typical current trace measured over the complex. The STM tip was fixed at the height corresponding to $V_s = 24$ mV and $I_t = 0.5$ nA during the measurement. The rapid flipping gives a wide feature in the high current state compared to the low current state. The abrupt drops of the current indicate the moment of the dissociation. In this case the complex is dissociated at the first current drop, and reformed the H bond at the abrupt increase of the current)

When a water molecule is reacted with a hydroxyl group locating on a neighboring row of the water, a different complex can be obtained (Figs. 9.6a–b). The product is characterized by an oval-shaped protrusion (label this product as [001]-complex) and the asterisk in Fig. 9.6b indicates the original position of the

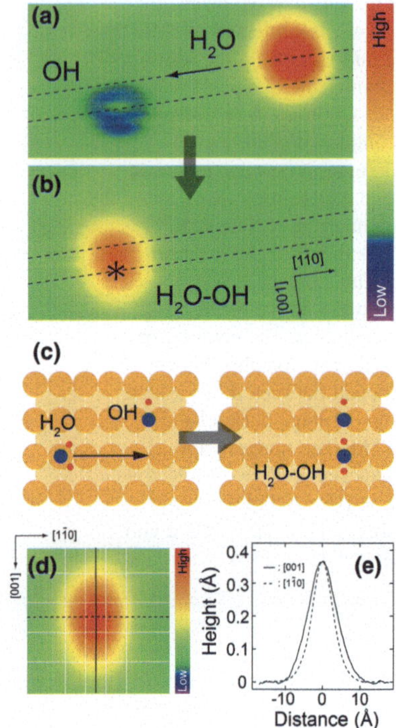

Fig. 9.6 **a** STM images of a water molecule and an OH group located on adjacent rows (*dashed lines*). **b** The complex produced under this geometry shows an oval shape. **c** Schematic illustration of the reaction. **d** An STM image of the complex on which the lattice of Cu(110) is superimposed. **e** The height profiles along the solid ([001]) and dashed ([1$\bar{1}$0]) lines in **d**. The sample bias voltage and tunneling current during the image acquisition were $V_s = 24$ mV and $I_t = 0.5$ nA, respectively, and the size is 47×22 Å^2 for **a** and **b**, and 17.5×17.5 Å^2 for **d**

hydroxyl. Thus two isomers of water-hydroxyl complex can be produced selectively depending on the initial geometries of the reactants. No interconversion between the two complexes was observed. The enlarged image of a [001]-complex with the lattice of Cu(110) is shown in Fig. 9.6d. As shown in Fig. 9.6e, the height profiles along the [001] (solid line) and [1$\bar{1}$0] (dashed) directions clearly indicate C_{2v} symmetry of the appearance.

The dissociation of a [001]-complex into a water and hydroxyl requires a voltage pulse higher than 0.5 V, suggesting the larger binding energy than that of a [1$\bar{1}$0]-complex. On the other hand, below 0.5 V, a [001]-complex shows a hopping along the atomic row direction. Using the atom tracking routine, the average hopping rate was investigated. A typical trace measured at $V_s = 453$ mV is shown in Fig. 9.7a, where the displacement of the [001] (red line) and [1$\bar{1}$0] (blue line) directions is plotted as a function of the measurement time. This result clearly indicates the hopping occurs exclusively along the [1$\bar{1}$0] direction. The time intervals between the hopping events were collected by repeating the tracking routine, and the average hopping rates were obtained. Figure 9.7b shows the voltage dependence of the rate, where the thresholds are observed around $V_s = 440$ and 330 mV for an H_2O-OH and D_2O-OD, respectively. The energy and isotope ratio of $\sim$1.3 suggests the excitation of the O–H(D) stretch mode [ν(OH(D))] triggers the hopping. The threshold energy is comparable to the ν(OH) of a free water molecule

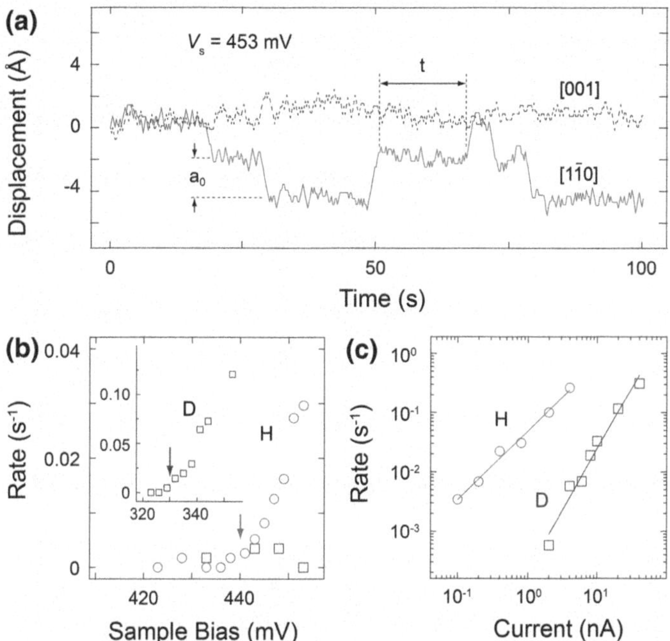

Fig. 9.7 **a** Typical trajectory of the STM tip when it tracks a [001]-complex (H_2O-OH) at $V_s = 453$ mV and $I_t = 0.5$ nA. The red and blue lines indicate the displacements along the $[1\bar{1}0]$ and [001] directions, respectively. **b** The hopping rates of a [001]-complexes as a function of bias voltage. The red circles and blue squares represent the rates for an H_2O-OH and D_2O-OD complexes, respectively. The tunneling current was fixed at 0.5 nA during the measurements. The inset shows the result for a D_2O-OD complex with the tunneling current of 20 nA. **c** The current dependence of the hopping rate measured at $V_s = 443$ and 343 mV for an H_2O-OH and D_2O-OD complexes, respectively. The reaction orders are 1.1 ± 0.1 and 2.0 ± 0.1 for an H_2O-OH and D_2O-OD complexes, respectively, indicating the hopping is induced via one- and two-electron processes

(457 meV[23]), suggesting that the free O–H stretching is associated with the process. It is noted that the deuterated complex is immobile at $I_t = 0.5$ nA even at $V_s = 450$ mV and it requires a relatively high current ($\sim$ 20 nA) to induce the hopping. The current dependence of the rate (Fig. 9.7c) indicates that the hopping is induced via one- and two-electron processes for H_2O-OH and D_2O-OD complexes, respectively. The hopping is induced through anharmonic coupling of the ν(OH) with the complex-substrate hindered translation mode. The higher reaction order for a D_2O-OD presumably suggests that the overtone excitation of ν(OD) ($\sim$660 meV) is required to be overcome the hopping barrier.

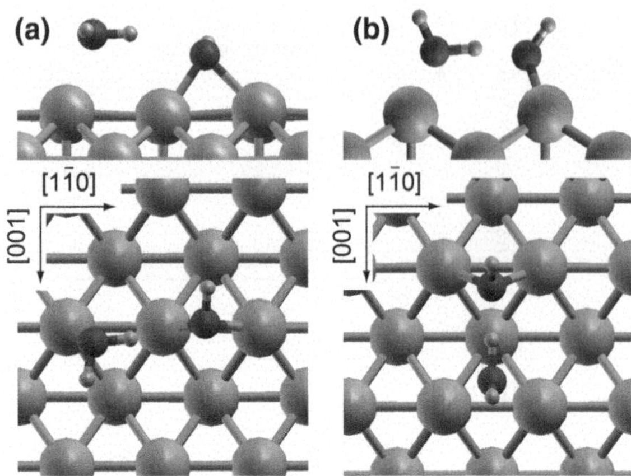

Fig. 9.8 The calculated structures for H_2O-OH complexes. (*upper*) Side and (*bottom*) top views. **a** [1$\bar{1}$0]-complex. **b** [001]-complex

9.2.2 Structure of Water-Hydroxyl Complexes on Cu(110)

The structure of water-hydroxyl complexes was determined by DFT calculations.[1] Two possible initial geometries are considered; first the water molecule is put on the same atomic row as the hydroxyl as an initial structure. Second the water molecule is put on the adjacent atomic row of the hydroxyl. Figure 9.8 shows the optimized structures. For the first case (Fig. 9.8a) the structure is stabilized by 0.13 eV compared to the isolated H_2O and OH on Cu(110), which corresponds to the energy of H bond. The water molecule is deviated from the on-top site, which is energetically favorable for a water molecule, to form the optimal H bond with the OH group. A [1$\bar{1}$0]-complex is assigned to this side-on structure. On the other hand, for the second case (Fig. 9.8b) the more stable structure can be obtained, where the complex is stabilized by 0.44 eV despite the water is fully displaced from the on-top configuration and both of the water and hydroxyl are bounded at the adjacent short-bridge sites along the [001] direction. The H bond in this structure is remarkably strong compared to that of a water dimer on Cu(110) which has a normal strength of H bond (0.14 eV). Since the adsorption energy of a water

[1] DFT calculations were performed using the STATE code [Y. Morikawa et al. Phys. Rev. B 69, 041403 (2004).] The calculations were conducted within the Perdew-Burke-Ernzerhof (PBE) generalized gradient approximation [J. P. Perdew et al. Phys. Rev. Lett. 77, 3865 (1996).]. PBE slightly overestimates the binding energies of H_2O dimer, and slightly underestimates the proton transfer barrier at a short distance ($\sim$2.5 Å), but is sufficiently accurate for the present purpose. The surface was modeled by a five-layer Cu slab with an H_2O-OH complex aligned along the [001] ([1$\bar{1}$0]) direction in a 3×3 (2×4) periodicity, and a 4×4 *k*-point set was used to sample the Brillouin zone. The adsorbates were put on one side of the slab, and the spurious electrostatic interaction was eliminated by the effective screening medium method [M. Otani and O. Sugino, Phys. Rev. B 73, 115407 (2006); I. Hamada et al. ibid. 80, 165411 (2009).]. Adsorbates and the topmost two Cu layers were allowed to relax, while remaining Cu atoms are fixed at their respective bulk positions.

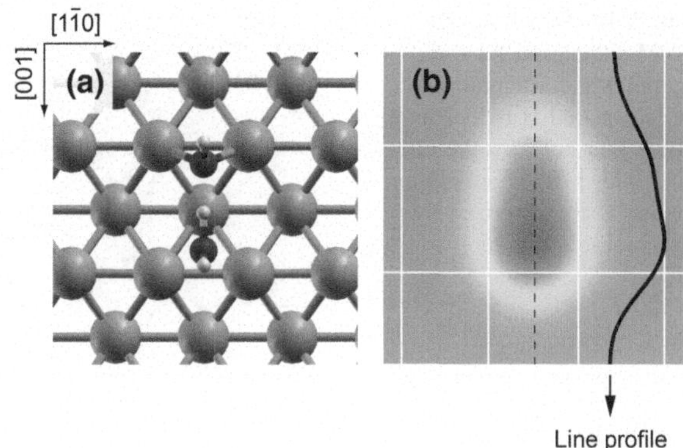

Fig. 9.9 STM simulation for the asymmetric configuration. The shared H makes a covalent bond with the one O atom in (**a**). **b** STM simulation image is characterized by asymmetric character, in which the maximum protrusion can be observed over the water unit

molecule isolated on Cu(110) was calculated to be 0.34 eV, a water molecule in the complex is stabilized totally by as large as 0.78 eV with respect to that in gas phase. A [001]-complex is assigned to this structure. This structure is, however, incompatible with the STM appearance of C_{2v}. Indeed, the simulated STM image for this structure (Fig. 9.9b) has a maximum protrusion over the water molecule. Thus the asymmetric character is clearly inconsistent with the experimental result.

9.2.3 Formation of a Symmetric Hydrogen Bond

To solve the contradiction between the experiment and calculations for a [001]-complex, I postulated that the shared H in a [001]-complex is located at the center between two O atoms due to the formation of a low-barrier H bond, yielding a symmetric configuration as shown in Figs. 9.10a. It is well known that the hydroxide ion (OH^-) is an excellent H bond acceptor and forms a strong H bond with water molecule [24, 25]. In fact, the calculations predict that a [001]-complex is characterized by the short $d_{O\text{-}O}$ (2.5 Å) and the strong H bond (0.44 eV). These results imply that the formation of a low-barrier H bond is plausible in a [001]-complex. The STM image simulated for this symmetric structure (Fig. 9.10b) shows a symmetric character, being very consistent with the experiment. Since the ZPE of nuclear is not included in the present calculations, where nuclear is treated as a classical particle, the structure of Fig. 9.8b represents classically favorable structure. Thus the inclusion of the ZPE possibly gives a different result.

To examine the speculation that the symmetric state (Figs. 9.10a) is preferred to the asymmetric one (Fig. 9.8b), we first calculated the adiabatic potential energy surface of the shared H when it is transferred between two O atoms. The adiabatic potential energy is shown in Fig. 9.11 (circles). At each point along the path, the positions of O atoms, residue H atoms and the topmost two Cu layers were

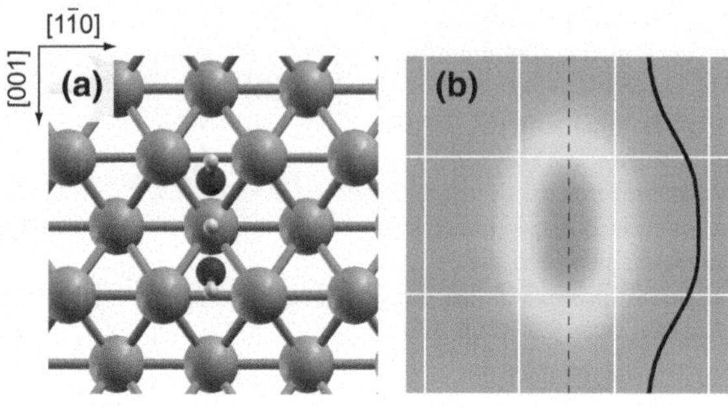

Fig. 9.10 Structure and STM simulation for the symmetric configuration. **a** The shared H is located at the center between the two O atoms. **b** STM simulation image is characterized by symmetric character

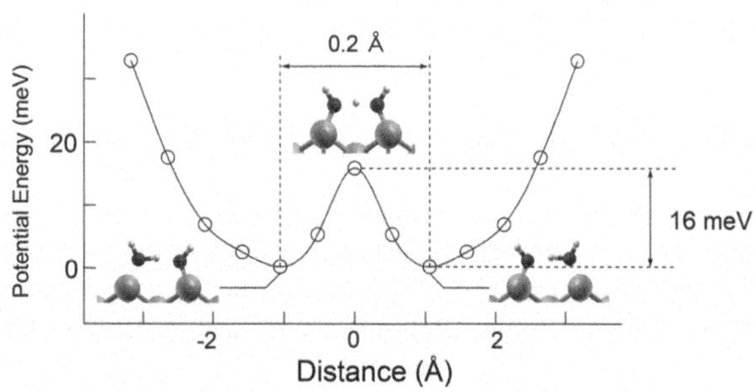

Fig. 9.11 Adiabatic potential energy for a [001]-complex along the minimum-energy path of the H transfer between the two O atoms. The two minima and peak correspond to the asymmetric and symmetric configurations, respectively

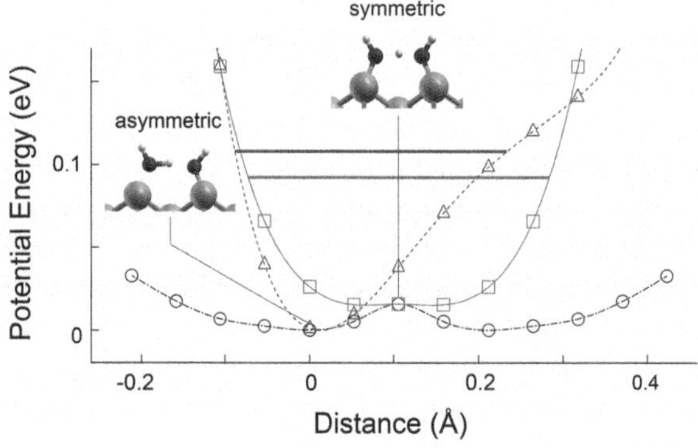

Fig. 9.12 Static potential energies for the symmetric (*red squares*) and asymmetric (*blue triangles*) configurations calculated by moving the H along the minimum-energy paths with other degrees of freedom fixed at their optimized geometries. Adiabatic potential is simultaneously plotted (*black circles*). The curves are obtained by a cubic spline fit to the calculated data. The dashed lines indicate the levels of total energies for each configuration

optimized. The potential shows a double-well structure with a significantly reduced barrier (16 meV) and small distance (0.2 Å) between the minima. Therefore the asymmetric configuration is slightly favored within the accuracy of the present DFT calculation. However, this result implies that the zero-point motion causes a preference of the symmetric configuration. The ZPE of the asymmetric and symmetric configurations is calculated, and compared their total

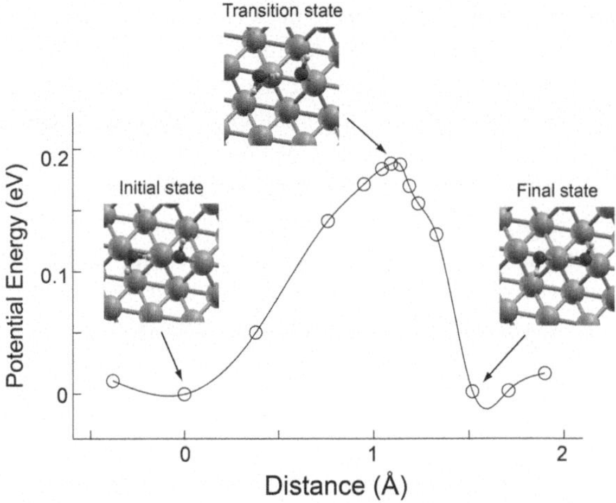

Fig. 9.13 Adiabatic potential energy for the [1$\bar{1}$0]-complex along the minimum-energy path of proton transfer between O atoms. The curve was obtained by a cubic spline fit to the calculated data. As the proton transfers, the binding site of the original OH group changes gradually from the bridge site to the top site, and that of the water molecule in the opposite way. The transition state lies in the off-centered position along the path, resulting in the potential curvature of asymmetric shape

energies (Fig. 9.12). Since H atom is expected to move much faster than heavy O and Cu atoms, the ZPE associated with the H atom transfer coordinate can be calculated by determining the static PES, where the degrees of freedom except for the shared H atom are frozen and the one-dimensional Schrödinger equations is solved. The calculated ZPEs are 107 and 76 meV for the asymmetric and symmetric configurations, respectively. The symmetric state has lower kinetic energy by ~30 meV and thus is stabilized in total energies (dashed lines). This suggests barrier less motion of the shared proton and thus formation of a symmetric H bond. The zero-point motion of the other degrees of freedom should be taken into account, but we assume the effect of their classical treatment is minor and negligible. I note that a D_2O-OD complex appeared almost similar to an H_2O-OH in the STM image, suggesting that it is still in the symmetric configuration, although the preference in the kinetic energy is reduced due to the doubled mass. As mentioned in introduction, the most suitable method is path integral ab initio calculation which can exactly take into account the ZPE of the system. However, it requires the massive platform for the calculation. Therefore we thought of ZPE in more feasible way.

For comparison, the adiabatic potential of a [1$\bar{1}$0]-complex was also calculated (Fig. 9.13). The transfer of the shared H requires the change in the adsorption site from the bridge (top) to top (bridge) for the original OH group (water molecule). Therefore, the potential barrier is relatively high (0.2 eV), suggesting that the side-on complex retains an asymmetric H bond (normal H bond). The present result has an implication to the water-hydroxyl complexes formed at elevated temperatures and higher coverage [26–32]. The water dissociation and hydroxyl formation on Cu(110) were investigated mainly by means of the x-ray photoemission spectroscopy. The dissociation is only partial and water molecules survive against desorption even at 428 K under a near-ambient pressure. It was proposed that water molecules are anchored to OH groups and stabilized against desorption due to a strong H_2O-OH interaction. The STM revealed that the water-hydroxyl chain complex grows along $[1\bar{1}0]$ after annealing the water-covered surface to $\sim$ 200 K. Since the water molecule is strongly bound in the bridge complex found in the present work ($E_b = 0.78$ eV), it may be a good candidate as a basis that composes the thermal products observed previously.

9.3 Summary

The hydration reaction between water and hydroxyl was controlled at the single molecule level using STM. Two distinct water-hydroxyl complexes were produced selectively and they can be controlled by the initial geometry of the reactants before the reaction. The structure and H-bonding nature in the complexes were investigated by the comprehensive DFT calculations. Upon the reaction between the two reactants on the same atomic row of Cu(110), the side-on complex was formed, which was identified by a relatively weak H bond (0.13 eV). On the other hand, upon the reaction between two reactants located on the adjacent rows, the bridge complex was formed, which was characterized by the strong H bond (0.44 eV) and the small distance between the two O atoms. Based on the detailed analysis of the STM appearance and adiabatic and static potential calculations, it is proposed that "a low-barrier H bond" is formed in the latter complex as a result of the significant reduction of effective barrier for the proton transfer and the ZPE within the H bond.

References

1. W.W. Cleland, M.M. Kreevoy, Science **264**, 1887 (1994)
2. P.A. Frey, S.A. Whitt, J.B. Tobin, Science **264**, 1927 (1994)
3. M.E. Tuckerman, D. Marx, M.L. Klein, M. Parrinello, Science **275**, 817 (1997)
4. M.E. Tuckerman, D. Marx, M. Parrinello, Nature **417**, 925 (2002)
5. P.W. Atkins, Physical Chemistry 6th end. 740-741 (Oxford Univ. Press, Oxford, 1998)
6. C.J.T. de Grotthuss, Ann. Chim. **58**, 5 (1806)
7. C. Reid, J. Chem. Phys.30, 182 (1959)
8. R. Pomès, B. Roux, J. Phys. Chem. **100**, 2519 (1996)
9. P. Loubeyre, R. LeToullec, E. Wolanin, M. Hanfland, D. Hausermann, **397**, 503 (Nature, London, 1999)
10. R. Ubbelohde, K.J. Gallagher, Acta Crystallogr. **8**, 71 (1975)
11. G. Held, D. Menzel, Phys. Rev. Lett. **74**, 4221 (1995)
12. G. Held, D. Menzel, Suf. Sci. **316**, 92 (1994)

13. A. Michaelides, P. Hu, J. Am. Chem. Soc. **123**, 4235 (2001)
14. P.J. Feibelman, Science **295**, 99 (2002)
15. S. Völkening, K. Bedürftig, K. Jacobi, J. Wintterlin, G. Ertl, Phys. Rev. Lett. **83**, 2672 (1999)
16. C. Clay, S. Haq, A. Hodgson, Phys. Rev. Lett. **92**, 046102 (2004)
17. T. Schiros, L.-Å. Näslund, K. Andersson, J. Gyllenpalm, G.S. Karlberg, M. Odelius, H. Ogasawara, L.G M. Pettersson, and A. Nilsson. J. Phys. Chem. C **111**, 15003 (2007)
18. L. Giordano, J. Goniakowski, J. Suzanne, Phys. Rev. Lett. **81**, 1271 (1998)
19. Y.D. Kim, R.M. Lynden-Bell, A. Alavi, J. Stulz, D.W. Goodman, Chem. Phys. Lett. **352**, 318 (2002)
20. B. Meyer, D. Marx, O. Dulub, U. Diebold, M. Kunat, D. Langenberg, C. Wöll, Angew. Chem. Int. Ed. **43**, 6641 (2004)
21. A. Michaelides, A. Alavi, D.A. King, Phys. Rev. B **69**, 113404 (2004)
22. X.-Z. Li, M.I.J. Probert, A. Alavi, A. Michaelides, Phys. Rev. Lett. **104**, 066102 (2010)
23. G. Herzberg, *Molecular Spectra and Molecular Structure I, Spectra of Diatomic Molecules* (Van Nostrand, Princeton, 1950)
24. P.A. Giguère, Rev. Chim. Miner. **20**, 588 (1983)
25. S.S. Xantheas, J. Am. Chem. Soc. **117**, 10373 (1995)
26. M. Polak, Surf. Sci. **321**, 249 (1994)
27. K. Bange, D.E. Grider, T.E. Madey, J.K. Sass, Surf. Sci. **137**, 38 (1984)
28. A. Spitzer, H. Lüth, Surf. Sci. **160**, 353 (1985)
29. Ch. Ammon, A. Bayer, H.-P. Steinrück, G. Held, Chem. Phys. Lett. **377**, 163 (2003)
30. K. Andersson, A. Gómez, C. Glover, D. Nordlund, H. Öström, T. Schiros, O. Takahashi, H. Ogasawara, L.G.M Pettersson and A. Nilsson. Surface Science Letters **585**, 183 (2005)
31. K. Andersson, G. Ketteler, H. Bluhm, S. Yamamoto, H. Ogasawara, L.G.M. Pettersson, M. Salmeron, A. Nilsson, J. Am. Chem. Soc. **130**, 2793 (2008)
32. J. Lee, D.C. Sorescu, K.D. Jordan, J.T. Yates Jr, J. Phys. Chem. C **112**, 17672 (2008)

Chapter 10
One-Dimensional Water-Hydroxyl Chain Complexes: Hydrogen-Atom Relay Reactions in Real Space

Abstract In this chapter I describe the real-space observation of H-atom relay reactions in one-dimensional (1-D) water-hydroxyl complexes where a water and hydroxyl molecules interact via H bond and are aligned along the [001] direction on a Cu(110) surface. The 1-D chain-like complexes, represented by H_2O–$(OH)_n$ ($n = 2$–4) are assembled in a fully controlled fashion using a combination of the dissociation and manipulation of individual water molecules. H-atom relay reactions are induced by an injection of energetic electrons from the STM tip. In the relay a water molecule that is initially located at the end of the chain is structurally transferred to the other end through the sequential H- and covalent bond exchange. The relay rate is investigated using the time-resolved measurement of STM. The voltage and current dependences of the rate clearly indicate that the vibrational excitations trigger the relay. The experimental findings are rationalized by ab initio calculations for the adsorption geometry, active vibration modes, and reaction pathway of the H-atom relay, which provide a microscopic insight of the elementary processes.

Keywords Hydrogen transfer reactions · One-dimensionally hydrogen-bonding system · Single molecule reactions through inelastic electron tunneling

10.1 Introduction

Assembling desirable nanoscale structures with the atomic-scale precision and controlling their functions are not only a dream for chemist but also key challenges towards developing molecule-based devices. As seen in the previous chapters, STM enables us to image, manipulate and characterize single atoms and molecules on conductive surfaces. Furthermore tunneling electrons from an STM tip can be used as the energy source to control adsorbate motions such as hopping, rotation,

T. Kumagai, *Visualization of Hydrogen-Bond Dynamics*, Springer Theses,
DOI: 10.1007/978-4-431-54156-1_10,

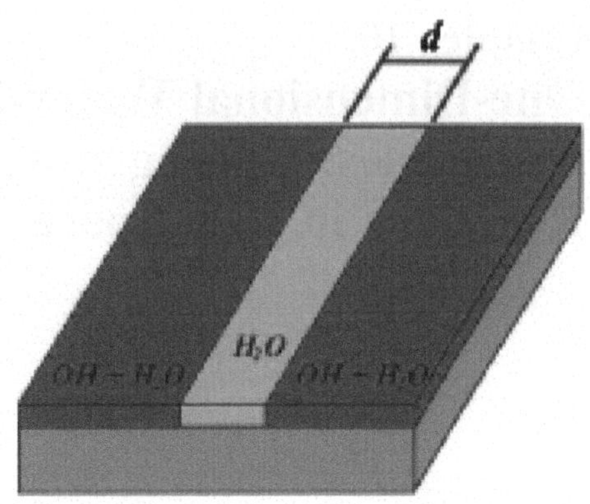

Fig. 10.1 Schematic of 1-D surface modulated distribution OH + $H_2O/H_2O/$ OH + H_2O on Pt(111). The width of H_2O region d was 400 μm. Reprinted with permission from Ref. [7]. Copyright 2008, American Physical Society

switching and desorption, as well as to initiate chemical reactions such as dissociation, bond formation, dehydrogenation and *cis–trans* conversion. STM has also been used to study sequential bond dissociation/formation reactions propagating along self-assembled molecular chains [1]. Here I show a new class of single-molecule chemistry with STM involving H bonds. By engineering a platform of H-bonded one-dimensional chains on a Cu(110) surface we are able to transfer an H-atom from one end of the chain to the other.

H-transfer reactions play a crucial role in many chemical and biological processes [2], yet still imperfectly understood, which is also related to many functional materials such as proton conductors [3], organic ferroelectric compounds [4], and confined liquids mimicking protein channels [5]. H-bond between water molecules is the path of H-transfer and hydronium ion is structurally-transferred from one location to the other through the sequential exchange of H- and covalent bond in condensed phase. This process is well-known as Grotthuss mechanism [6] and described in Chap. 9, which rationalize the anomalously high mobility of protons (H^+) in aqueous solutions compared to other cations. The elementary processes of H-transfer have been simulated by ab initio calculations, providing the detailed picture of the transportation mechanism. However the Grotthuss mechanism has never been directly observed in experiment. This is because it is extremely difficult to eliminate complex environmental effects in conventional spectroscopic methods that rely on ensemble from many species.

In surface chemistry, H-atom/proton transfer is of fundamental importance in the elementary processes of electrode chemistry, heterogeneous catalysis and energy productions. The experimental characterization is still extremely limited. Recently, H-atom migration on a metal surface via the relay mechanism was investigated using a combination of the laser-induced thermal desorption (LITD) and the micro-meter scale x-ray photoelectron spectroscopy (micro-XPS) [7]. The unique approach to observe the H-transfer is illustrated in Fig. 10.1. First 2-D H-bonding network composed of H_2O and OH is formed on Pt(111) and then the H_2O–OH layer is patterned by the LITD using a pulsed laser beam focused into small spot, yielding the narrow bare area on the Pt surface. Then the patterned

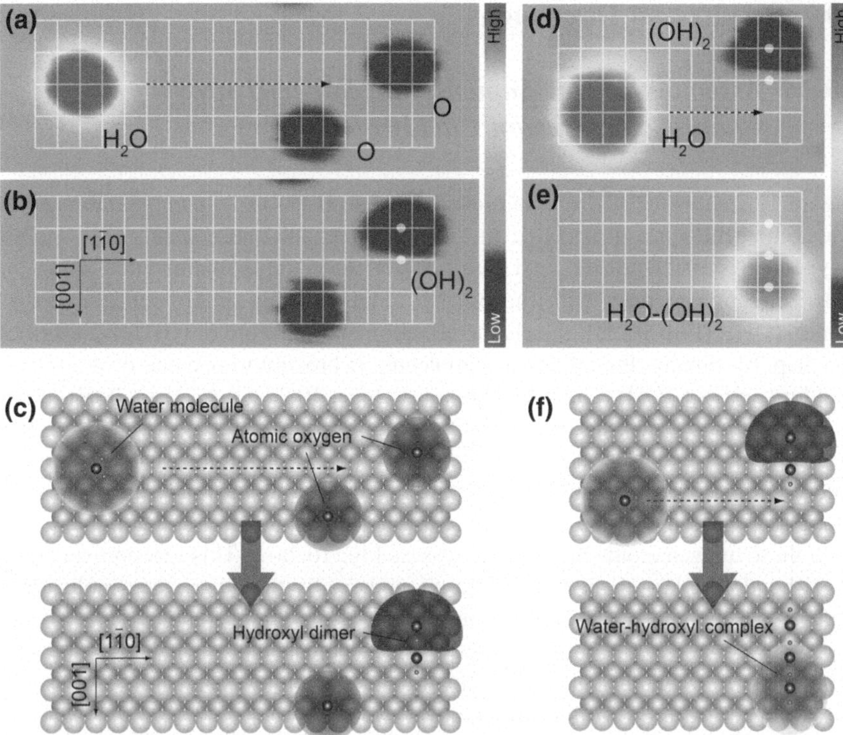

Fig. 10.2 Sequential STM images of the assembly process of an H_2O–$(OH)_2$ complex from individual water molecules on a Cu(110) surface. **a** A water molecule and two oxygen atoms are imaged as a round protrusion and two depressions, respectively. The white grid lines indicate the lattice of Cu(110). The water molecule is dragged along the [1$\bar{1}$0] direction toward atomic oxygen (*dashed arrow*). **b** The reactants yield an $(OH)_2$ imaged as a semicircular depression. The white dots indicate the nearest short-bridge sites to which oxygen atoms in $(OH)_2$ are bonded. **c** Schematic illustration of the process of (**a–b**). (**d–e**) Subsequently, the other water molecule is manipulated toward the end of $(OH)_2$ (*dashed arrow*), then yielding an H_2O–$(OH)_2$ complex. The three dots in (**e**) depict the nearest short-bridge sites to which oxygen atoms in the complex are bonded. It is noted that the positions of oxygen atoms deviate slightly from the exact short-bridge sites along [001] axis, as shown in the structure optimized by DFT calculations (Fig. 10.3a and b). **f** Schematic illustration of the process of (**d–e**). The images were recorded in constant current mode at $V_s = 24$ mV and $I_t = 0.5$ nA. The image sizes are 20×58 Å^2 for (**a** and **b**), and 20×36 Å^2 for (d) and (e)

H_2O–OH overlayer is exposed to H_2O to recover the bare Pt area with water. The time-evolution of the distribution of H_2O and OH along the surface was observed by the micro-XPS. From statistical analyses based on a diffusion equation, they found two distinguishable proton transfer pathways with the diffusion coefficient of $\sim 3.0 \times 10^{-12}$ and $\sim 9.9 \times 10^{-13}$ m^{2}s^{-1}, and dwell time of ~ 5.2 and 48 ns for the simple proton transfer from H_2O to adjacent OH at 140 K.

10.2 Results and Discussions

10.2.1 Assembly of One-Dimensional Water-Hydroxyl Chain Complexes on Cu(110)

Figure 10.2 show the assembling procedure of an H_2O–$(OH)_2$ complex. First some water molecules are dissociated into OH species and subsequently into atomic oxygen (Fig. 10.2a). In the next step a water molecule is manipulated along the $[1\bar{1}0]$ direction to react with the oxygen atom, yielding an H-bonded hydroxyl dimer, $(OH)_2$ (Fig. 10.2b); the characterization and structure of $(OH)_2$ is described in Chap. 8]. Finally, the other water molecule is brought to one end of the $(OH)_2$, yielding an H_2O–$(OH)_2$ chain-like complex (Figs. 10.2d–f). An H_2O–$(OH)_3$ chain is assembled by moving an H_2O–OH complex to an $(OH)_2$ (Fig. 10.3) [the structure of H_2O–OH is described in Chap. 9]. An H_2O–$(OH)_4$ is assembled as follows; First, two oxygen atoms separating by $2b_0$ along [001] direction each other are prepared (Fig. 10.4a). Then two water molecules are sequentially reacted with these oxygen atoms in the same way as Fig. 10.2a–b. This reaction generates an H-bonded $(OH)_4$ chain (Fig. 10.4b–c), and finally the other water molecule is attached to the end of the $(OH)_4$, yielding an H_2O–$(OH)_4$ complex (Fig. 10.4d). Every reaction described above takes place spontaneously when the reactants come close to each other. It is noted the produced complexes are sufficiently stable to be imaged with a low bias voltage.

In addition to 1-D chain configurations, two structural isomers of H_2O–$(OH)_2$ can be obtained depending on the initial position of a water molecule that reacts with an $(OH)_2$. A water molecule is attached to a hydroxyl that acts as the H-bond acceptor and donor in Figs. 10.5a–b and c–d, respectively. Because hydroxyl is expected to be a strong H-bond acceptor, I propose the structures of the side-on complexes where the water molecule donates its H bond to the hydroxyl group in $(OH)_2$ for both cases (the tentative model is illustrated on the bottom of STM images of Fig. 10.5b and d). A high voltage pulse ($\sim$5 V) can transform these side-on complexes into the chain configuration via several unidentified intermediate states.

10.2.2 Structure of an H_2O–$(OH)_2$ Chain Complex on Cu(110)

The structure of H_2O–$(OH)_2$ was determined by DFT calculations.[1] Figure 10.6a shows the optimized structure, where the water molecule and hydroxyl species are

[1] The DFT calculations were carried out with a plane-wave basis set and ultrasoft pesudopotential method [D. Vanderbilt, Phys. Rev. B 41, 7892 (1990).] as implemented in the STATE code [Y. Morikawa et al. Phys. Rev. B 69, 041403(R) (2004).]. Perdew-Burke-Ernzerhof (PBE) [J.P. Perdew et al. Phys. Rev. Lett. 77, 3865 (1996).] generalized gradient approximation (GGA) was used for the exchange–correlation functional. Plane-wave cutoffs of 36 Ry and 400 Ry were

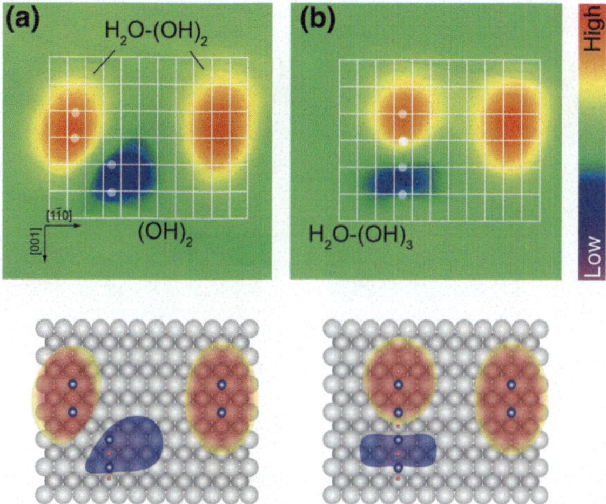

Fig. 10.3 a–b Sequential STM images and schematic illustration of the assembly process of an H_2O–$(OH)_3$ complex from $(OH)_2$ and H_2O–OH (38 × 38 Å^2, V_s = 30 mV and I_t = 0.5 nA). The white grid lines indicate the lattice of the Cu(110) surface and the dots represent the short-bridge sites occupied by oxygen atoms. The H_2O–OH complex is moved along the [1$\bar{1}$0] direction (*dashed arrow*) with the STM, and reacted with the $(OH)_2$ to yield an H_2O–$(OH)_3$ complex

bounded at the short-bridge site. It is noted that the water molecule and its neighboring hydroxyl are strongly interacted via H bond and they are slightly deviated from the exact short-bridge position. The corresponding simulated STM image[2] (Fig. 10.6b) is characterized by a "tadpole"-shaped protrusion with the head and tail appeared over the water molecule and hydroxyl groups, respectively, being in good agreement with the experimental appearance.

(Footnote 1 continued)
used to expand wave functions and augmentation charge, respectively. The Cu(110) surface was modeled by a five-layer slab with a 4 × 3 periodicity, and a vacuum region of 12.89 Å was inserted between slabs. The slab was constructed using the lattice constant optimized by PBE-GGA (a_0 = 3.64 Å). An H_2O–$(OH)_2$ complex was put on one side of the slab and spurious electrostatic interaction was eliminated by the effective screening medium method [I. Hamada et al. Phys. Rev. B 80, 165411 (2009).]. Adsorbate and topmost two Cu layers were allowed to relax until the forces were less than 0.05 eV/Å, and remaining Cu atoms are fixed at their respective bulk positions. Brillouine zone sampling was done using a 4 × 2 Monkhorst–Pack *k*-point set [H.J. Monkhorst and J.D. Pack, Phys. Rev. B 13, 5188 (1976).] and Fermi-surface was treated by the first-order Methfessel-Paxton scheme [A. Methfessel and A.T. Paxton, Phys. Rev. B 40, 3616 (1989).] with the smearing width of 0.05 eV.

[2] For the STM simulations, a dense 16 × 8 *k*-point set was used. Adsorbate and topmost two Cu layers were allowed to relax, and remaining Cu atoms are fixed at their respective positions. STM simulations were conducted within the Tersoff-Hamman theory [J. Tersoff and D.R. Hamann, Phys. Rev. Lett. 20, 1998 (1983).]. In the STM simulations, the sample bias voltage was set to 25 mV, and images were obtained at the constant height of 7 Å from the topmost Cu plane.

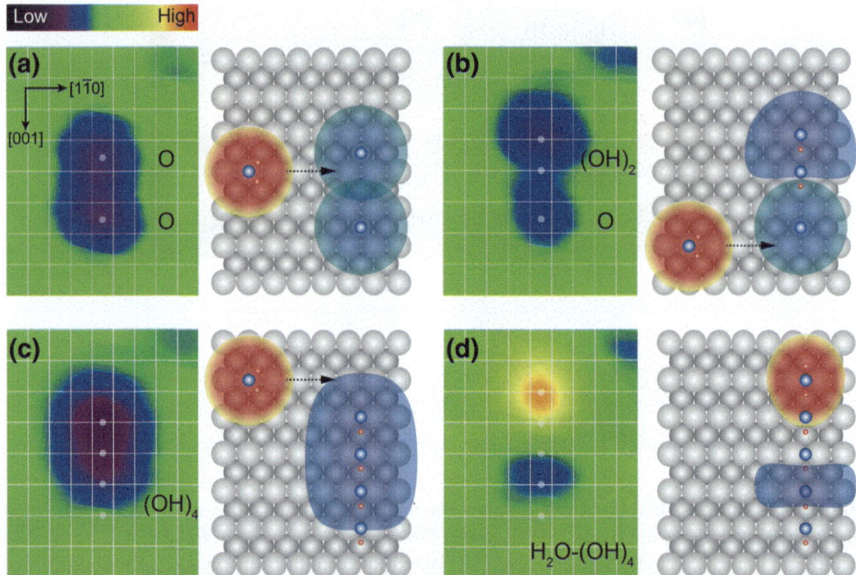

Fig. 10.4 Sequential STM images during the assembly process of an H_2O–$(OH)_4$ complex from two oxygen atoms and three water molecules, along with the schematic illustration of the reactions. **a** Two oxygen atoms are adsorbed along the [001] direction with the interatomic distance of $2b_0$. The two white dots indicate the hollow sites to which the oxygen atoms are adsorbed. **b** A water molecule is manipulated toward one oxygen atom to yield $(OH)_2$ + O. **c** A second water molecule reacted with another oxygen atom to yield $(OH)_4$. **d** Finally, a third water molecule was attached to the end of $(OH)_4$, resulting in a H_2O–$(OH)_4$ chain complex. The dots in **c** and (**d**) indicate the short-bridge sites occupied by oxygen atoms, although they are displaced from the exact short-bridge sites. The images were recorded at $V_s = 30$ mV and $I_t = 0.2$ nA for (**a–c**) and $I_t = 0.5$ nA for (**d**). The image size is (23 × 29 Å^2). The color bar corresponds to the topographic height from −0.46 Å (*low*) to 0.35 Å (*high*)

10.2.3 Direct Observation of Hydrogen-Atom Relay Reactions

A voltage pulse with the STM over a water molecule induces the flip of the chain-complex. The flip can be observed in a series of the chain complexes (Fig. 10.7b–d). By comparison to the calculated structure and the STM simulation, it is obvious that the flip corresponds to the H-atom relay reaction and the water molecule is structurally-transferred from the end of the chain to another (Fig. 10.7a). According to the detailed analysis of the STM image, the binding site of the oxygen atoms in the complex is not altered before and after the flip, suggesting the migration of the constituent molecules to the other site is not included. Thus only the H-atom relay is merely involved in the process.

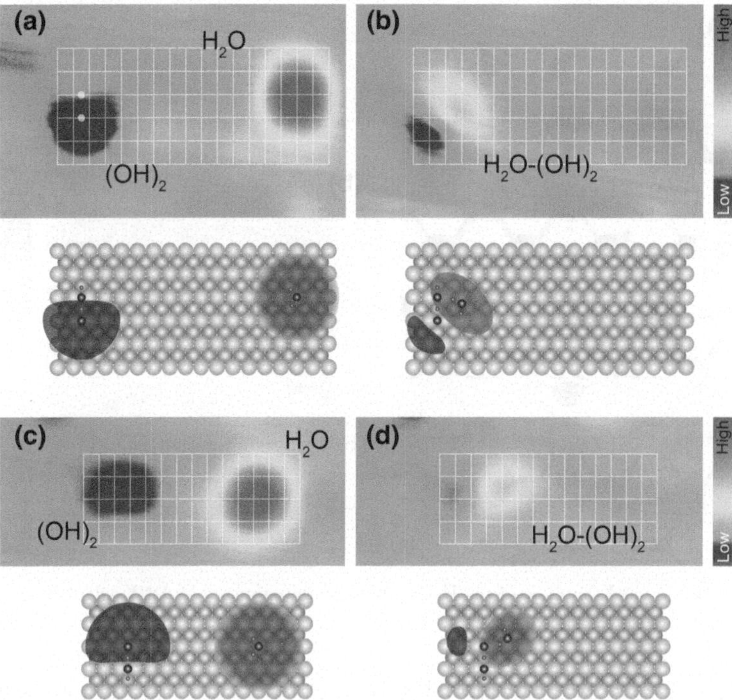

Fig. 10.5 Sequential STM images during the assembly process of two side-on H_2O–$(OH)_2$ complexes with schematic illustrations. The *white grid* lines in the STM images indicate the lattice of a Cu(110) surface and white dots depict the positions of oxygen atoms. The complexes in (**b**) and (**d**) arose from the reaction in (**a** and **c**), where water molecule reacted with the H-bond acceptor and donor in an $(OH)_2$, respectively. The model structure of the side-on complexes is schematically shown in the *bottom* of (**b** and **d**). The size is 33 × 55 Å^2 for (**a** and **b**), and 24 × 57 Å^2 for (**c** and **d**). The images were recorded at $V_s = 24$ mV and $I_t = 0.5$ nA

10.2.4 Quantitative Analysis of Hydrogen-Atom Relay Reactions

The H-atom relay reaction is reversible and can be directly observed by monitoring the tunneling current with the open feedback measurement. The inset of Fig. 10.8 shows a typical current trace measured over a water molecule (detailed parameters is described in the figure caption). The high and low current states correspond to the initial and final configurations of the relay, respectively, and the current jumps are the moments of the individual relay events. The relay rate is determined by a statistical analysis of the time interval between the relay events. No other state without the two states is observed within the STM time-resolution, suggesting the transition state like OH-H_2O-OH is not stable and its lifetime is very short.

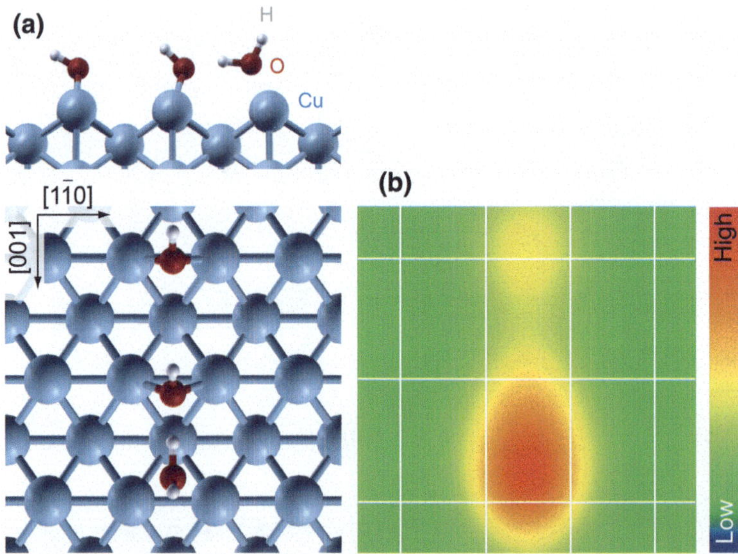

Fig. 10.6 **a** Side and top view of the H_2O–$(OH)_2$ chain optimized by DFT calculations. **b** STM simulation for (**a**)

To explore the mechanism of the H-atom relay, the bias voltage, tunneling current, and spatial dependence of the relay rate is investigated for an H_2O–$(OH)_2$, D_2O–$(OD)_2$, and H_2O–$(OH)_3$. Figure 10.8a shows the quantum yield [reaction probability per tunneling electron] as a function of bias voltage. The STM tip is positioned over the water molecule during the measurements. The yield are observable at the range from 10^{-12} to 10^{-6} per electron within the experimental time scale. The lowest limit is determined by a realistic time scale of experimental tolerance and the highest limit by the bandwidth of a preamplifier for STM. For H_2O–$(OH)_2$ the yield shows an initial increase around 180 mV, a moderate increase beyond ~220 mV, and a sharp enhancement around 430 mV. For D_2O–$(OD)_2$ obvious isotope effects can be observed, where the yield shows an initial increase around 200 mV and a sharp enhancement around 320 mV. Figure 10.8b shows the current dependence of the relay rate. The rate shows a linear dependence onto the current at several voltages, indicating that the transfer is induced via one-electron process over the whole bias range. I confirmed that the same results are obtained with inversed the bias polarity, ruling out the possibility that the electric field induces the reaction. Above results unambiguously suggest that the relay is triggered by the vibrational excitation of the adsorbate molecules. On the other hand, the reaction yield of H_2O–$(OH)_3$ (Fig. 10.8a, green triangles) is found to be much smaller than that of H_2O–$(OH)_2$ by several orders of magnitude. Figure 10.8c shows the spatial dependence of the relay yield at $V_s = 243$ and 443 mV, where the yield is plotted as a function of the distance along the H-transfer axis (the [001] direction) as indicated in the inset. At both voltages the

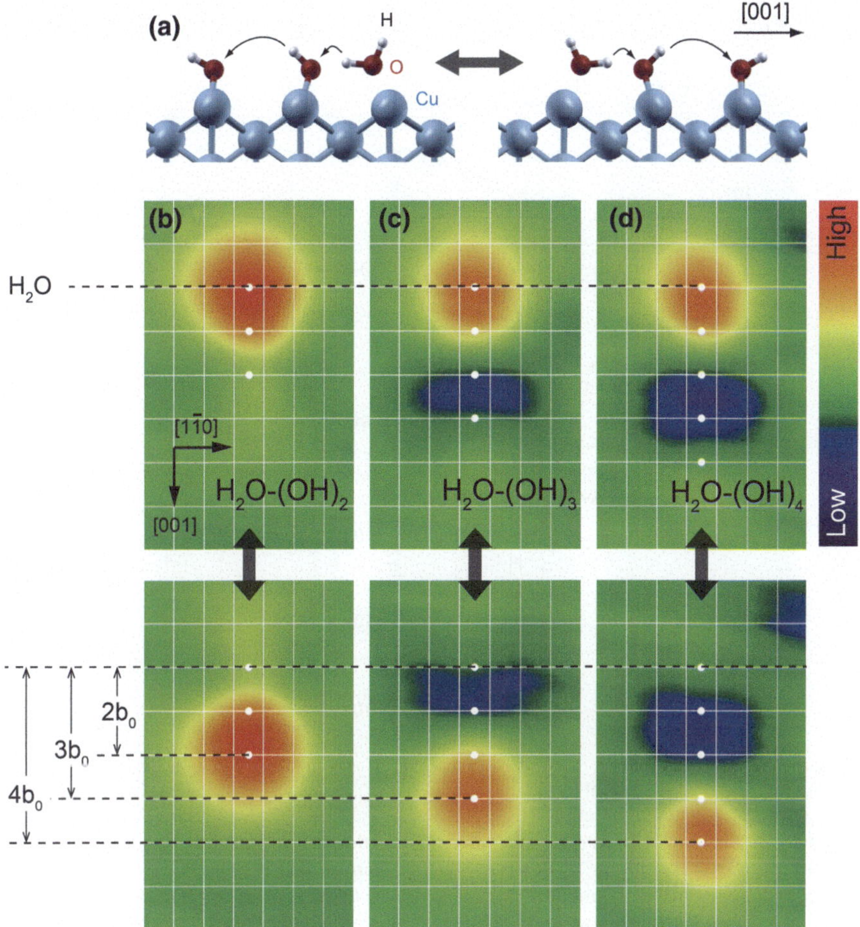

Fig. 10.7 **a** Side view of the H_2O–$(OH)_2$ chain optimized by DFT calculations **b** STM images of the H_2O–$(OH)_2$ chain and its counterpart superimposed with the lattice of Cu(110) (*white lines*). The appearance can be inverted by the voltage pulse of STM over the protrusion. The inversion was also observed for (**c**) H_2O–$(OH)_3$ and (**d**) H_2O–$(OH)_4$ chains. The dots in (**b–d**) indicate the nearest short-bridge sites binding oxygen atoms in the chains (note that oxygen atom positions deviate from the exact short-bridge sites as shown in Fig. 10.6a). The inversion of the appearance corresponds to H-atom relay reaction in which a sequential H-atom transfer is included as shown by curved arrows in (**a**).The images were obtained at $V_s = 24$ mV and $I_t = 0.5$ nA (18×29 Å^2). The range of the height shown in the color scale is from −0.36 to 0.44 Å

yield is the largest when the tip is positioned over the water molecule, but at $V_s = 443$ mV it is more broadly distributed over the chain, indicating that different vibrational modes are involved at these different voltages.

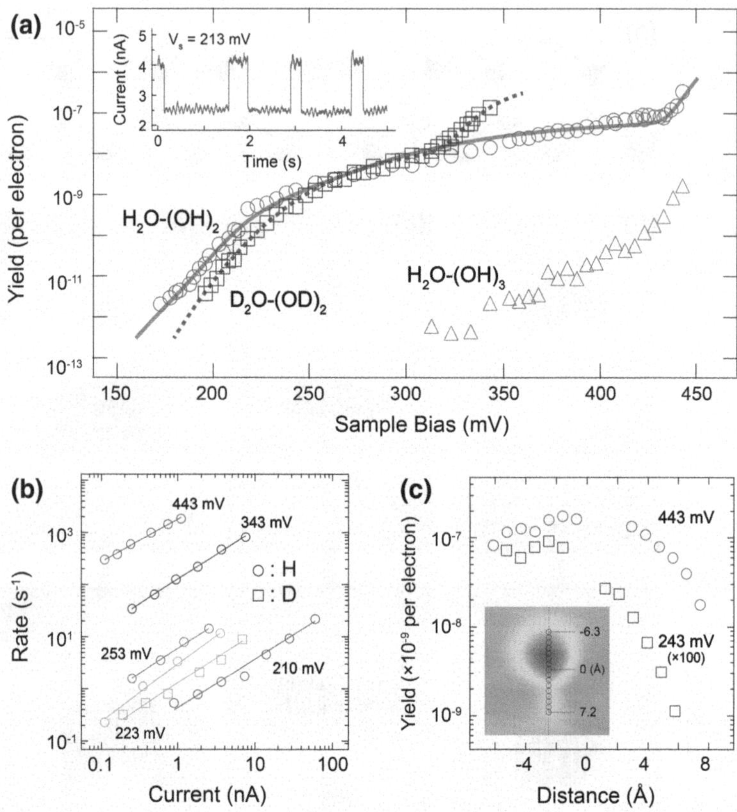

Fig. 10.8 **a** Yields of H-atom relay reactions as a function of the applied bias voltage measured for an H_2O–$(OH)_2$ (*red circles*), D_2O–$(OD)_2$ (*blue squares*), and H_2O–$(OH)_3$ (*green triangles*). The inset shows a typical current trace recorded with the tip fixed over the protrusion for an H_2O–$(OH)_2$ at $V_s = 213$ mV. The two-state fluctuation indicates the back-and-forth relay reaction in the chain. The *red* and *blue curves* are Eq.10.10 applied to an H_2O–$(OH)_2$ and D_2O–$(OD)_2$ data, respectively, with the parameters in Table 10.1. **b** The current dependence of the reaction rates for an H_2O–$(OH)_2$ (*red circles*) and D_2O–$(OD)_2$ (*blue squares*) at various voltages. The slopes were estimated to be 0.82 ± 0.03, 0.95 ± 0.02, 0.97 ± 0.04, and 0.93 ± 0.05 for $V_s = 443$, 343, 223 and 210 mV, respectively. The data are presented in logarithmic scale and the slopes are close to unity, indicating single-electron processes in the whole voltage range. **c** Position dependence of the yield along the chain for an H_2O–$(OH)_2$ at $V_s = 243$ mV (*blue squares*) and $V_s = 443$ mV (*red circles*).The position is represented by the distance from the center of the inversion along [001], as shown in the inset with the STM image. While the yield is broadly distributed over the complex at $V_s = 443$ mV, it is more localized at $V_s = 243$ mV above the protrusion (water molecule)

10.2.5 Theory Describing Inelastic Electron Tunneling Action Spectroscopy

Following Refs. [8, 9] and references herein, I describe a theory to quantify the relation between the reaction yield and vibrational excitation. The purpose of the following formulation is to reproduce the voltage dependence of the reaction rate (yield), namely STM-AS, where the vibrational excitation drives the reaction or motion of adsorbates via the indirect process (anharmonic coupling between the high frequency mode and reaction coordinate modes). The outcome is applicable to the systems which have an essentially same mechanism. As already mentioned, the reaction yield $Y(V)$ is given by.

$$Y(V) = \frac{R(V)}{I_{\mathrm{tot}}(V) \cdot e} \tag{10.1}$$

where $R(V)$ is the reaction rate and $I_{\mathrm{tot}}(V)$ is the total tunneling current and e is the elementary electric charge. The definition of power law of $R(V)$ gives

$$R(V) = k(I_{\mathrm{in}}(V))^n \tag{10.2}$$

with k being the rate constant, I_{in} being the inelastic tunneling current, and n being the reaction order, namely the number of tunneling electron required to induce reactions or motions of an adsorbate. Here $I_{\mathrm{tot}} = I_{\mathrm{el}} + I_{\mathrm{in}}$ (I_{el} is the elastic tunneling current) and the Iin vanishes if the energy of the tunneling electron, eV, is lower than a specific vibrational energy $\hbar\Omega$ whereas it increases linearly above $\hbar\Omega$. Here I consider only the case of $n = 1$ (one-electron process) and assume the indirect process where the reaction coordinate mode is indirectly excited via the anharmonic coupling of the high frequency mode. In the framework of IET process in which the tunneling electron induces the molecular vibration via resonance scattering, I_{in} is proportionate to the vibrational generation rate $I_{\mathrm{in}} = k'\Gamma_{\mathrm{iet}}$, Thus.

$$R(V) = K\Gamma_{\mathrm{iet}}(V) \tag{10.3}$$

where K is the prefactor that is determined by the elementary process like electron–hole damping rate and anharmonic coupling between the high frequency and reaction coordinates. Given a vibration mode characterized by a certain vibrational density of state $\rho_{\mathrm{ph}}(\omega)$ and define the vibrational generation rate in terms of a spectral representation, Γ_{iet} can be written by.

$$\Gamma_{\mathrm{iet}}(V) = \int_0^{\infty} d\omega \rho_{\mathrm{ph}}(\Omega)\Gamma_{\mathrm{in}}(\omega, V) \tag{10.4}$$

where $\Gamma_{\mathrm{in}}(\Omega, V)$ is the spectral generation rate corresponding to an excitation of energy $\hbar\Omega$. The spectral generation rate is given by.

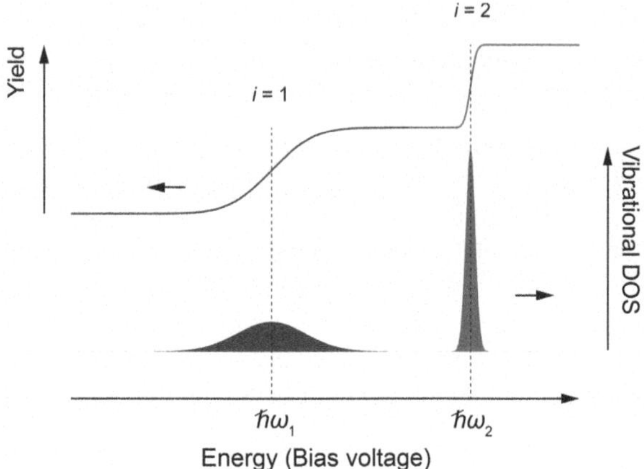

Fig. 10.9 Schematic simulation of this Eq. 10.10. The broader vibrational DOS causes more moderate increase in the reaction yield

$$\Gamma_{\text{in}}(\omega) = 2\pi\chi^2 \frac{\Gamma_s\Gamma_t}{(\Gamma_s+\Gamma_t)^2} \int d\varepsilon\rho_a(\varepsilon)\rho_a(\varepsilon+\omega)[n_s(\varepsilon+\omega) - n_t(\varepsilon+\omega)][n_s(\varepsilon) - n_t(\varepsilon)] \tag{10.5}$$

$n_{s/t}(\varepsilon)$ and $\Gamma_{s/t}$ are the Fermi-distribution function and reactant tunneling coupling constant of the substrate/tip. At low temperature limit, namely $k_BT \ll \hbar\Omega$, Eqs. 10.4 and 10.5 lead to.

$$\Gamma_{\text{iet}}(\omega, V) \approx \gamma_{\text{eh}}(\omega)\frac{\Gamma_{\text{t}}}{\Gamma_{\text{s}}}\frac{|eV| - \hbar\omega}{\hbar\omega}\theta(|eV| - \hbar\omega) \tag{10.6}$$

where Γ_s and Γ_t are the tunneling couplings (wide-band limit) to substrate and tip sides, respectively. And we assume the tunneling condition of $\Gamma_t \ll \Gamma_s$. Here $\gamma_{\text{eh}}(\omega) \approx 2\pi\chi^2\rho_{\text{a}}(E_{\text{F}})$ is the electron–hole pair damping of the vibration and $\theta(x)$ is the Heaviside step function. Written $I_{\text{tot}} = \sigma_0 V$ in terms of the junction conductance σ_0, consequently.

$$Y(V) = K_{\text{eff}}\frac{1}{V}\int_0^{|eV|} d\omega\rho_{\text{ph}}(\omega)(|eV| - \hbar\omega) \tag{10.7}$$

can be obtained with the effective prefactor K_{eff}, which is written by.

$$K_{\text{eff}} = \frac{4\pi}{\hbar\sigma_0}\chi^2\rho_a(\varepsilon_F)^2\frac{\Gamma_t}{\Gamma_s}K \tag{10.8}$$

where χ is the electron-vibration coupling and $\rho_a(\varepsilon_{\text{F}})$ is the adsorbate density of states at the Fermi energy ε_{F}. Here we assume the Gaussian distribution for the vibrational density of state around a certain threshold energy Ω, thus.

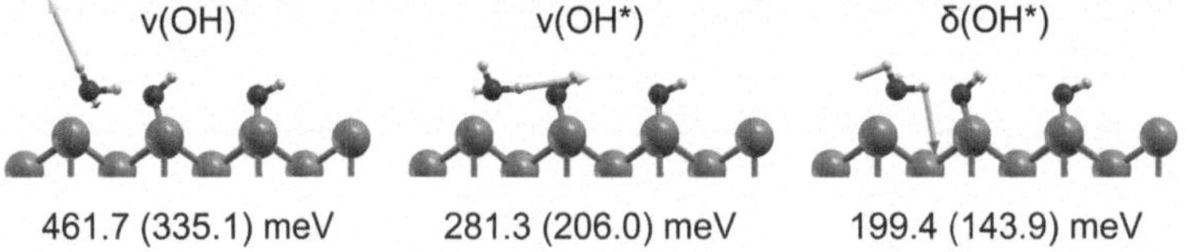

Fig. 10.10 Calculated vibrational mode for an H_2O–$(OH)_2$ and D_2O–$(OH)_2$ on a Cu(110) surface

$$\rho_{\mathrm{ph}}(\Omega, \sigma_{\mathrm{ph}}; \omega) = \frac{1}{\sigma_{\mathrm{ph}}\sqrt{2\pi}} \exp\left(\frac{(|eV| - \Omega)^2}{2\sigma_{\mathrm{ph}}^2}\right) \tag{10.9}$$

The motivatiobn for a Gaussian distribution is to take into account all broadening effects in a single parameter ρ_{ph}, in particular the nature of the broad spectrum of the OH (OD) stretch mode in a single hydrogen bond, but also other effects such as temperature, intrinsic life-time broadening, and instrumental ones [10]. Finally, when several, distinct vibration modes are involved in the process, the total reaction yield should be expressed as the sum over the accepting modes contributions:

$$Y(V) = \frac{1}{V}\sum_i K_{\mathrm{eff}}^i \int_0^{|eV|} \rho_{\mathrm{ph}}^i(\Omega_i, \sigma_{\mathrm{ph}}^i; \omega)(|eV| - \hbar\omega) \tag{10.10}$$

Here the index i represents the specific vibration mode. Figure 10.9 is schematic expression of Eq. 10.10. We can find a broad (narrow) distribution of the vibrational density of state (DOS) results in a moderate (rapid) increase in the reaction yield.

10.2.6 The Mechanism of Hydrogen-Atom Relay Reactions

As shown by the solid curves in Fig. 10.8a, Eq. 10.10 reproduces the experimental data using the parameters given in the Table 10.1. These parameters suggest that three characteristic vibrational modes (free OH/OD stretch, OH*/OD* stretch, and H_2O scissors, where H*/D* denotes the shared H/D atom in the H bond) are involved in the relay reaction in the range of voltages where the transfer yield could be observed. The observed onset around 430 [320] mV for an H_2O–$(OH)_2$ [D_2O–$(OD)_2$] can be associated with the free OH [OD] stretch mode. At lower voltages we assign modes centered at 310 and 262 meV for H_2O–$(OH)_2$ and D_2O–$(OD)_2$ to the shared OH*/OD* stretch modes. In the case of H_2O–$(OH)_2$ a third mode around 213 meV is taken into account and associated with the H_2O scissors mode. On the other hand, for D_2O–$(OD)_2$ the scissors mode is well below the voltage range where the yield is observable. It is noted that the OH*/OD* stretch modes are significantly red-shifted from the free OH/OD stretch and also

Table 10.1 The parameters determined by the spectral fitting of the voltage dependence of the relay yield

Mode	ħΩ (meV)	σ_{ph} (meV)	e^2K_{eff}
$\delta(H_2O)$	215	12	9×10^{-7}
$\nu(OH^*)$	305	35	7×10^{-5}
$\nu(OH)$	454	10	5×10^{-3}
$\delta(D_2O)$	–	–	–
$\nu(OD^*)$	261	22	1.5×10^{-5}
$\nu(OD)$	333	10	1.3×10^{-3}

characterized by very large broadenings (38 meV for OH* and 23 meV for OD*). The significant mode softening with respect to the free stretch modes and spectacular enhancement of the width are familiar from IR spectroscopy and known to originate from the strong anharmonic character of a single H bond [11]. Both anharmonicity of the potential as well as a strong coupling to low-energy intermonomer modes affect the detailed features. Furthermore, since anharmonicity is expected to be more pronounced for H than D, the model parameters for OH*/OD* are also consistent with the largest shift and width in the case of OH*.

In order to rationalize the fit parameters vibrational energies of an H_2O–$(OH)_2$ and D_2O–$(OD)_2$ complex were calculated by DFT. The results are summarized in Fig. 10.10, which is very consistent with the fitting result.

The vibrational mode analysis also explains why the spatial distribution of the transfer yield (Fig. 10.8c) is more localized over the water molecule at $V_s = 243$ mV than at $V_s = 443$ mV. In the former case only the H_2O scissors, which is expected to be spatially localized at the water molecule within the chain, can be excited. In the latter case, however, also the OH*/OH stretch modes of both the water molecule and hydroxyl groups within the chain can be excited to trigger the transfer, eventually resulting in a broader distribution.

The reaction pathway was also investigated by total energy calculations[3](Fig. 10.11). The initial step is transportation of the shared H-atom to the center hydroxyl, which is almost barrier-less ($\sim$0.04 eV). This is in line with the recent DFT calculations that predicted the H-atom sharing in the water and hydroxyl overlayers is facile due to the zero-point energy of the nuclear motion [12]. The subsequent H-bond cleavage between the OH and the center water molecule constitutes the highest barrier of 0.25 eV, in which the displacement of the center water molecule along [001] direction is mainly involved. The OH, OH* stretch and

[3] In order to estimate the reaction barrier the nudged elastic band (NEB) method [G. Mills et al. Surf. Sci. 324, 305 (1995).] was used with 12 floating images between the initial and final configuration. After convergence of the 12 images, the transition state was determined by another NEB calculation with just one floating image between the two images closest to the top of the barrier. The one-dimensional potential landscape is reported in Fig. 10.11 along with the corresponding geometries as insets. The force tolerance for the NEB calculation was set to 0.05 eV/Å. The reported total energy differences were evaluated using the tetrahedron method with Blöchl corrections using a Γ-centered mesh of 16 × 8 *k*-points.

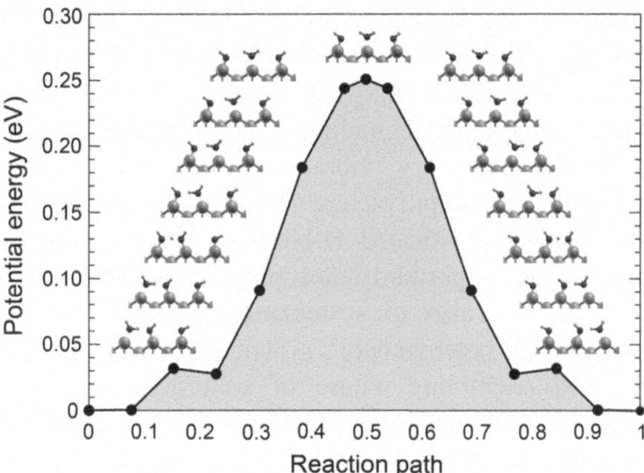

Fig. 10.11 Calculated potential energy surface for the H-atom transfer reaction of H_2O–$(OH)_2$ on a Cu(110) surface. The H-atom moves along [001] direction. The set of images from initial to final states, shown as insets, were determined by the nudged elastic band method. The H-transfer to the center OH is facile (~0.04 eV). The transition state (0.25 eV) corresponds to OH–H_2O–OH with a broken H bond

H_2O scissors modes are therefore postulated to couple to the reaction coordinate for the H-bond cleavage. It is noted that the H-atom transfer is observed below the bias voltages of 250 mV in the bias voltage dependence (Fig. 10.8a), where tunneling electron should not have enough energy to induce the H-atom transfer via the over-barrier process. Although it is still open question, I propose that the H-bond cleavage, which mainly involves the rotational motion of a center water molecule, might here proceed via vibrationally-assisted tunneling [9] of an entire H_2O or D_2O molecule through the transition state. The transfer reaction is therefore completed either via tunneling (at low bias) or over-barrier processes (at high bias). For longer chains [H_2O–$(OH)_{3,4}$], the H-atom transfer reaction occurs in sequence and thus the energy dissipation is more significant, giving rise to lower reaction yields than for an H_2O–$(OH)_2$. Such energy dissipation plays a crucial role in determining the product, yield, and pathway in chemical reactions on surface.

10.3 Summary

H-atom relay reactions were visualized with one-dimensional H-bonded water-hydroxyl complexes assembled with the single molecule precision using STM. The H-atom relay reaction is induced by vibrational excitations, where the water molecule situated at the end of the complex is *structurally transferred* to another end via the multiple H-/covalent bond exchange. The voltage, current, and spatial dependence of the relay rate (yield) unveiled the behind mechanism. In the voltage dependence, the increase of the relay yield was observed at vibrational energies of the scissors, shared OH, and free OH stretch modes with the rational isotope shift. The current dependence revealed the relay is driven by one-electron processes in the whole voltage region. Furthermore the spatial dependence of the yield showed the vibrational excitation is localized over a water molecule at the energy of the scissors mode excitation, while it becomes broad at that of the OH stretch mode due to the contribution from hydroxyls. I also found the relay yield is significantly reduced due to the energy dissipation during the relay as the chain is lengthened. These experimental findings were rationalized by ab initio calculations for adsorption geometry, vibration mode, and reaction pathway, which provide a detailed microscopic picture of the elementary processes. Engineering even longer and more sophisticated H-bonding systems supported on solid surfaces could provide an opportunity not only to achieve mechanical logic circuits using H-atoms but also to systematically study fundamental steps of the H-atom dynamics in heterogeneous systems. Such systems are conceivable by combining the self-assembling nature of water/hydroxyl complexes [13–15] with STM manipulation techniques. Our discovery that the H-transfer along H-bonds is possible directly on metal surfaces suggests that the relay reactions may occur more generally at metal–molecule interfaces and therefore in liquids all the way down to the confining surfaces. This is of importance in diverse fields such as nanofluidics and design of hybrid materials for proton conduction.

References

1. P. Maksymovych, D.C. Sorescu, K.D. Jordan, J.T. Yates Jr, Science **322**, 1664 (2008)
2. J.T. Hynes, J.P. Klinman, H.-H. Limbach, R.L. Schowen, *Hydrogen-Transfer Reactions* (Wiley–VCH, Weinheim, 2007)
3. S. Bureekaew, S. Horike, M. Higuchi, M. Mizuno, T. Kawamura, D. Tanaka, N. Yanai, S. Kitagawa, Nature Mater. **8**, 831 (2009)
4. S. Horiuchi, Y. Tokura, Nature Mater. **7**, 357 (2008)
5. C. Duan, A. Majumdar, Nature Nanotech. **5**, 848 (2010)
6. C.J.T. de Grotthuss, Ann. Chim. **58**, 54 (1806)
7. M. Nagasaka, H. Kondoh, K. Amemiya, T. Ohta, Y. Iwasawa, Phys. Rev. Lett. **100**, 106101 (2008)
8. S.G. Tikhodeev, H. Ueba, Phys. Rev. B **70**, 125414 (2004)
9. S.G. Tikhodeev, H. Ueba, Phys. Rev. Lett. **102**, 246101 (2009)
10. K. Motobayashi, Y. Kim, H. Ueba, M. Kawai, Phys. Rev. Lett. **105**, 076101 (2010)
11. Y. Maréchal, The hydrogen bond and the water molecule. (Elsevier, Oxford, 2007)
12. X.-Z. Li, M.I.J. Probert, A. Alavi, A. Michaelides, Phys. Rev. Lett. **104**, 066102 (2010)
13. T. Yamada, S. Tamamori, H. Okuyama, T. Aruga, Phys. Rev. Lett. **96**, 036105 (2006)
14. J. Lee, D.C. Sorescu, K.D. Jordan, J.T. Yates Jr, J. Phys. Chem. C **112**, 17672 (2008)
15. J. Carrasco, A. Michaelides, M. Forster, S. Haq, R. Raval, A. Hodgson, Nature materials **8**, 427 (2009)

Chapter 11
Conclusions

I described the real space observation of H-bond dynamics within water-based model systems assembled on a metal surface. Low-temperature scanning tunneling microscope was employed to image, assemble and characterize such systems at the single molecule level. The combination of manipulation and controlled dissociation of individual water molecules made it possible to construct a variety of H-bonded systems. In addition to the assembly of desirable model systems, the time-resolved measurement of STM was proven to be useful to unveil the process of H-bond dynamics at the single molecule level. The voltage and current dependence of a motion or a reaction provided the key to elucidate the elementary processes. Especially, the voltage dependence is directly associated with vibrational excitations of molecule and quite useful for chemical identifications of adsorbates as well as for determining dynamical properties.

This research shed a new light on understanding of H-bond dynamics at a heterogeneous system in terms of the single molecule limit. In particular, visualization of quantum dynamics, such as H-bond interchange tunneling in a water dimer, single-proton flip motion in hydroxyl group and a symmetric H-bond in a water-hydroxyl complex, is highlighted. Furthermore, H-atom relay reactions, like the Grotthuss mechanism, were directly observed in one-dimensional H-bonded complexes for the first time. However, the time-resolution of STM was still limited to capture the transition state of the reaction process. To improve the time-resolved imaging would pave a novel way to elucidate detailed processes of H-bond dynamics. Additionally, the investigation of the temperature dependence of those processes gives abundant information of potential landscape determining the dynamical properties.Water and hydroxyl spontaneously form a wide variety of ordered structures on material surfaces. The combination of this self-assembling nature with STM manipulation gives a way to realize much more sophisticated and complex systems to examine H(-bond) dynamics.

T. Kumagai, *Visualization of Hydrogen-Bond Dynamics*, Springer Theses, DOI: 10.1007/978-4-431-54156-1_11,

The manufacturer's authorised representative in the EU is Springer Nature Customer Service Centre GmbH, Europaplatz 3, 69115 Heidelberg, Germany. If you have any concerns regarding our products, please contact ProductSafety@springernature.com

Printed and bound by CPI Group (UK) Ltd, Croydon, CR0 4YY

15/07/2026

02167645-0006